Next Generation Networks

Next Generation Networks
Perspectives and Potentials

Dr Jingming Li Salina
LiSalina Consulting, Switzerland
Pascal Salina
Swisscom SA, Switzerland

John Wiley & Sons, Ltd

Other Wiley Editorial Offices

John Wiley & Sons Inc., 111 River Street, Hoboken, NJ 07030, USA

Jossey-Bass, 989 Market Street, San Francisco, CA 94103-1741, USA

Wiley-VCH Verlag GmbH, Boschstr. 12, D-69469 Weinheim, Germany

John Wiley & Sons Australia Ltd, 42 McDougall Street, Milton, Queensland 4064, Australia

John Wiley & Sons (Asia) Pte Ltd, 2 Clementi Loop #02-01, Jin Xing Distripark, Singapore 129809

John Wiley & Sons Canada Ltd, 6045 Freemont Blvd, Mississauga, ONT, L5R 4J3, Canada

Wiley also publishes its books in a variety of electronic formats. Some content that appears in print
may not be available in electronic books.

Anniversary Logo Design: Richard J. Pacifico

Library of Congress Cataloging in Publication Data

Salina, Jingming Li.
   Next Generation Networks : perspectives and potentials / Jingming Li Salina and Pascal Salina.
     p. cm.
   Includes index.
   ISBN 978-0-470-51649-2 (cloth)
   1. Telecommunication systems. 2. Convergence (Telecommunication) 3. Business planning.
   I. Salina, Pascal. II. Title. III. Title: NGN perspectives and potentials.
   TK5102.5.S28 2007
   004.6—dc22

                          2007033010

British Library Cataloguing in Publication Data

A catalogue record for this book is available from the British Library

ISBN 978-0-470-51649-2 (HB)

Typeset in 10/12pt Times by Integra Software Services Pvt. Ltd, Pondicherry, India
Printed and bound in Great Britain by TJ International Ltd, Padstow, Cornwall
This book is printed on acid-free paper responsibly manufactured from sustainable forestry in
which at least two trees are planted for each one used for paper production.

To our parents

Contents

List of Tables

List of Illustrations

Preface

Right after experiencing the wonder of mobile communications, we are the lucky generation witnessing the latest miracle, 'the birth of Next Generation Networks – NGN'. NGN is today a very busy field that enjoys a wild pace of development. NGN standardization is going on in parallel within different international bodies while many telecom operators already claim that they are implementing NGN. Institutes and universities are still conducting research while, at the same time, products are starting to be available on the market.

However, NGN remains one of the most used buzzwords in the worlds of telecommunications, Internet and broadcasting. Working in this field and, after attending many workshops, conferences and thoroughly studying of the topic, the authors are convinced that a clear and comprehensive vision of NGN is still missing, a vision which, in authors' opinion, is essential for the strategic development of NGN.

Having absorbed the quintessence of the ITU-T pioneer work on NGN and other studies and with the benefits of many years of R&D and strategic working experience, the authors would like to provide their vision to the readers. Fitting the existing pieces together in an overview, this vision should deliver a complete and comprehensive picture of NGN.

Sometimes also known as pervasive computing or ambient intelligence, the future of ICT can be described from many points of view. This book is about NGN, its potentials and perspectives; it is about networks linking humans, devices and computers.

The emphasis is conceptual, with one eye on today's networks and another on the networks of the next generation. The intention is to envision the NGN and to explain it, to guide operators in the design of their network evolutionary path and vendors in developing their products and, last but not least, to inspire the researcher's or reader's creativity to catalyse the development of NGN.

Following an old Chinese proverb, advising one to 'cast a brick in order to attract jade', the authors welcome discussion about NGN and about the views expressed in this book, with the sincere hope of raising awareness about NGN to a higher level.

Acknowledgements

We would like to thank those organizations that granted us the permission to use and reproduce their graphical or text materials: ITU, ETSI, TMF and WWF international.

We also owe very much to Dr Albert Kuhn, who has been working for environmental and corporate responsibility at Swisscom for many years and has contributed tremendously in both fields. He has generously allowed us to use the material he has developed alone or with colleagues. We also appreciate very much his suggestions relating to Chapter 9.

We appreciate the artistic work of Michèle Mouche, who designed the illustrations and pictograms of Chapter 2.

We would like to thank our friend and colleague Mohamed Mokdad, at Swisscom, for his friendship and support with the standards.

We would like to thank our parents for their unlimited support and encouragement.

Last but not least, we would like to specially thank Mark Hammond, Rowan January, Sarah Hinton, Katharine Unwin, Wendy Pillar, Brett Wells and Vidya Vijayan at John Wiley & Sons, for their encouragements and professional support.

1

Introduction

Today, traditional telecom operators are facing two life-and-death challenges brought about by competition and technology:

- Challenge 1: Internet Service Providers (ISP) providing communication services based on the Internet Protocol (IP) are replacing traditional telecom services, not only at lower costs but also with enhanced features and almost unlimited potential.
- Challenge 2: overwhelming technology development on the one hand complicates the infrastructure investment decision but on the other hand enables competition as more players enter the telecom market.

However, nothing will happen overnight. There are ways for operators to survive these challenges: challenge 1 – the telecom operator should *become more than an ISP*; and challenge 2 – the telecom operator should *apply a model of operation driven by the needs of customers*.

1.1 CHALLENGE 1: TO BECOME MORE THAN AN ISP

ISP provide various IP-based communication services that are software-enabled and which they can therefore deploy rapidly and widely in various ways. However, as ISP services are delivered to customers over the physical networks of incumbent telecom operators, the quality of service delivered or the service experience perceived by the customer relies heavily on the performance of the underlying transport network. Therefore, ISP cannot fully guarantee the service quality delivered to customers.

In contrast, a telecom operator has the control of service quality in its hands and can seize this advantage to become more than an ISP. For some time traditional telecom operators have tried to provide ISP services; however, in spite of following the mainstream ISPs, the innovative spirit of the latter has not yet been incorporated by operators. Subject to willpower and effort, it will be only a question of time before telecom operators catch up with ISP in terms of IP-based services.

The way for a telecom operator to outrun an ISP is:

- to master the performance control of its underlying network, which goes beyond keeping the network working, as happens today;
- to be more innovative and aggressive than the ISP in creating and delivering IP-based services.

1.2 CHALLENGE 2: TO APPLY A MODEL OF OPERATION DRIVEN BY CUSTOMER NEEDS

The telecommunications industry has entered a new era. The level of technology development exceeds the level of customer desire, as represented on Figure 1.1, and the difference between the two is increasing rapidly.

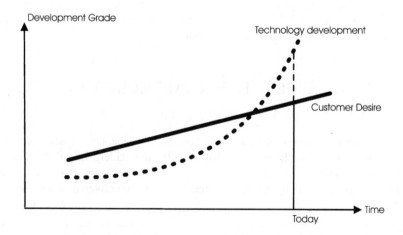

Figure 1.1   Technology development vs customer desire

The current boom in technology development expands the playground of telecom operators, but also makes its investment in infrastructure more complex. Furthermore, it empowers new players.

How to retain customers with adequate investment has become the central question an operator must answer, in order to stay at the front of this increasingly competitive market. This leads to a change in the operation model from a technology-driven to a customer-need-driven model.

1.2.1   The Technology-driven Operation Model

The technology-driven model is well suited in a period when the technical development lags behind customer needs, as indicated in Figure 1.1, before the crossing point. In this situation, a newly developed technology fulfils the customer's practical needs and can therefore be adopted immediately, even without marketing stimulation. A typical example is mobile telephony, where the feature 'mobility' serves the practical need of the general customer if not of every customer.

Under the technology-driven model, the focus is on the technology. The infrastructure vendors control the telecom industry in the following ways:

- Vendors guide and master the technology development and build the infrastructure with services for operators.
- Operators follow the development, introduce the technology when it is mature and keep it working.
- Customers subscribe to an operator and are served with services coming with the infrastructure; technically all customers are treated equally on a first-come, first-served basis.

In short, the technology-driven model puts the vendor in the driver's seat. The control chain Vendor → Operator → Customer describes the model.

The role of operator is rather simple and passive: buy an infrastructure and keep it working. The operator's strategy reduces to when and how fast to introduce a new technology. Owning the latest technology and setting prices as low as possible for the same technology drives the competition among operators.

1.2.2   The Operation Model Driven by Customer Needs

Thanks to a rapid and constant development of technology, the world of telecom has entered a new era, graphically located after the crossing point in Figure 1.1.

In this domain, the newly developed technologies are beyond the practical need of most customers and their adoption is very selective. A typical example is the overwhelming emergence of mobile and wireless access technologies such as UMTS, HSPA, LTE, WiMAX, Flash-OFDM and Wireless MeshNetwork.

Under these circumstances, continuing with the technology-driven model would mean unaffordable investments and nevertheless unavoidable customer losses. It is therefore advisable to switch to an operation model centred on the needs of customers, which we will call the customer-needs-driven model.

The customer-needs-driven model suggests that operators focus on the service value brought to the customers; for instance services which improve the quality of life or which bring a new life experience. Under the customer-needs-driven model, the operator is the driver of the telecom industry. The relationship Customer → Operator → Vendor describes this model.

This change of paradigm requires:

- a vendor to focus more on the product and solution development in an innovative, fast-to-market and cost-effective way, besides understanding the needs and requirements of the operator.
- an operator to make tremendous efforts to

  o constantly study the *practical and potential needs of its customers* in order to build up a clear business vision and strategy;
  o thoroughly understand *the essence of existing and emerging technologies* in order to make the right choice for the development of infrastructure;
  o satisfy customers in a differentiated way – this requires

    (a) a flexibility to add and remove services according to the actual needs of customers;
    (b) a real-time manageability to guarantee the service according to a service level agreement (SLA);
    (c) a customer self-care including on-line subscription, service provisioning, account checking, etc.

  o behave in a responsible way towards the society and the environment, taking its corporate responsibility seriously and thus building up trust and winning the heart of the customer;
  o set requirements for new products and new solutions for vendors.

In summary, under the technology-driven model, the fundamental questions to answer are what can be sold to customers based on installed or upcoming technology and how and when to sell.

Under the customer-needs-driven model, the fundamental questions for an operator are what the current and the potential needs of the customers are and which technology can or will enable an operator to satisfy the identified customer needs.

1.3 NGN – THE HOLY GRAIL FOR A TELECOM OPERATOR?

How can a traditional operator become more than an ISP and how can it operate with the customer-needs-driven model?

The concept of NGN (Next Generation Networks) has been initiated and designed for traditional telecom operators to become more than ISP and to operate with the customer-needs-driven model in order to survive the challenges mentioned in the first paragraph. It brings fundamental changes to service network architecture and service network management. Figure 1.2 illustrates the architecture of NGN.

Service Network Architecture

The NGN service network architecture is based on information and communication technology (ICT), where the information technology part (IT) is mainly devoted to creating and delivering service and the communication technology part (CT) is for transporting data.

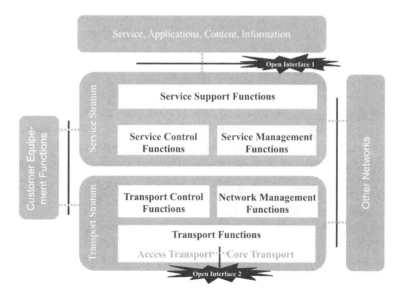

Figure 1.2 NGN service network architecture and management. Reproduced with the kind permission of ITU

The distinguishing characteristics of this service network architecture are:

- the separation of service layer and transport layer, thus enabling operators to add, upgrade and remove services without touching the transport network, and setting the conditions for an unlimited potential for IP-based services;
- the open interface between service creation and service delivery (indicated as open interface 1 in Figure 1.2) that enables a standardized way of creating services – this feature opens the door to many new providers, including consumers for advanced service, application, content and information;
- the open interface between the core and access network (indicated as open interface 2 in Figure 1.2) furthermore enables

  o the addition or the removal of an access network without changing the core network;
  o the provision of the same service through all kinds of access networks, although with a service quality that can be different.

- The customer equipment can be an end-user device, a home or vehicle LAN gateway, a corporate network gateway, a sensor or a machine.

Service Network Management

The NGN service network management is end-to-end, from customer management $\rightarrow$ service management $\rightarrow$ terminal management $\rightarrow$ network management $\rightarrow$ server farm management. The end-to-end management can be grouped into two parts, located on the service and the transport strata. The two parts communicate with each other in real time to fulfil real-time and customer-oriented management.

- The management on the service stratum is responsible for delivering services with a sufficient quality to end users by first setting up the performance requirements for the transport stratum and second monitoring the service quality received at the customer end. The performance requirements are set according to service priority, service quality requirements, customer subscription, actual customer SLA satisfaction level and the business relationship with the service provider.
- The management on the transport stratum is in charge of assigning adequate network resources for end-to-end transportation and guaranteeing the performance thereof. The network resource is allocated in an optimized manner according to the performance requirements from the service stratum, the network resource availability, the terminal capability, etc.
- A proactive management of the network performance is provided, where the performance data are measured along the end-to-end transport chain and analysed in real time, and action is generated when necessary as follows:

○ When a problem is detected, the relevant information is generated and delivered to the relevant teams, which are:

  (a) a management team, supplied with information on service impact and financial consequences;
  (b) a customer care team, requiring explanation and description of the problem, delay, repair time, etc.;
  (c) an operation team, that needs information on the cause of problem and the level of emergency to be resolved.

○ When a problem is predicted, a preventive measure should be activated to prevent the problem occurring.
○ When a customer has difficulty in running the requested service, adequate software is sent to the terminal.

- Terminal management is provided to enable the requested service using the correct configuration and adequate software.
- Third-party service management includes discovery, registration and use of the network services provided by an operator.

The NGN thus enables virtually unlimited IP-based services, far beyond an imagination that is certainly shaped and bound by today's available telecommunications, broadcasting and Internet services. Therefore, operators should not consider NGN simply as an upgrade or another novel network technology to follow. NGN represents a tremendous step ahead of today's networks and many new developments are still needed. However, the NGN concept and philosophy should be clear enough today for operators to orient their efforts and shape their network's evolution.

Today, several standards bodies, including ITU-T, ETSI, ATIS and CJK, actively standardize and specify the technical features of NGN. However, these on-going works address only the early features of NGN, which represent more an improvement of today's network than the fundamental changes that potentially can be included in the concept of NGN.

1.4 NGN AIMS AT IMPROVING LIFE QUALITY AND BRINGING NEW LIFE EXPERIENCE

NGN was designed as a solution for operators to take up the challenges of competition and technology – and ends up with the potential to bring much more to human beings than we can imagine today. By networking everybody and everything on the earth, below the surface of the seas and in space, NGN offers the possibility to improve quality of life and bring new experiences to people.

Sometimes known as 'pervasive computing' or 'ambient intelligence' when it focuses on the computing part of NGN, the vision of future applications of

ICT describes the network of sensors, processors and terminals that will become part of our daily life: 'A billion people interacting with a million e-businesses through a trillion interconnected intelligent devices' (Gerstner, IBM, 2000, on pervasive computing).

NGN will also network computing power, thereby empowering scientific research to advance technologies such as biologics, thus bringing further improvements in human life. Regarding human-to-human communication, NGN will provide virtual face-to-face communication, all-media and context-aware. Distance-independent and rich in resources (if properly designed), NGN will palliate diverse shortcomings such as the language barrier and perhaps more serious physical handicaps whenever and wherever necessary.

For human-to-machine communication, the user interface will be similar to human-to-human communication, including the use of speech, touch and sign language. For machine-to-machine communication, the connection is real-time, on-demand (ad-hoc) and autonomous. For sensor-to-server communication, situation-dependent instant communication takes place towards data centres for processing information and generating alarms and reactions. An example could be early warning systems for natural disasters such as earthquakes, hurricanes and tsunami. The NGN access will be ubiquitous and quality (bandwidth) available on demand.

Considering the stage of today's networks, NGN will need to take time to evolve. It is a long-term process that will take place stepwise.

1.5   THE NETWORK EVOLUTION TOWARDS NGN

Thanks to the pioneer work of ITU and other standardization bodies, the NGN functional architecture is today mature and clear enough to guide network evolution. Early steps to make would be:

- To open the interface between application creation and delivery. Until now, most operators have controlled the service creation and delivery interface in order to protect their revenue. However, the future demands on IP services, collectively known as 'electronic applications' or 'e-applications' and including e-government, e-learning, e-health, e-banking and e-tourism, are beyond most current operators' capabilities. The open interface will enable third parties to develop and delivery such applications.
- To decouple the service layer from the network layer. Until now, networks have been built for a specific service or a specific service was fitted onto the network. Typical examples are voice and short messages services (SMS), the preliminary services built in the GSM network. The future ICT enabled by NGN will rely on IT for the IP-based services and on CT for the data transportation.

- To separate the access network and the core network (a common core, access agnostic). Until today, a core network has supported its own access network; the simplest example is the GSM access network supported by its own core network. Recently, UMTS core networks have also started to support GSM access. This concept will be extended within NGN. In NGN, a single core network may have multiple access networks and access networks can be added, upgraded and removed without impacting on the core network.
- To manage an end-to-end system driven by customers needs. Until now, the management system has comprised a business supporting system (BSS) and an operation supporting system (OSS), where:

  o the BSS is responsible for everything related to customers – billing, customer care, customer relationship management (CRM);
  o the OSS is responsible for everything related to technology, such as network operation, fault management.

Operations centred on customers will require a communication between the BSS and the OSS in order to have the BSS direct the OSS action.

1.6 THE TELECOM ENVIRONMENT AND CORPORATE RESPONSIBILITY

The NGN brings hopes of a better future but has undesirable side effects that need to be considered when planning or implementing NGN features. Assessments of the technology, using tools such as life cycle analysis (LCA) or risk analysis, have identified the negative as well as the positive effects of NGN.

Direct impacts on the environment such as energy and material consumption, waste management or non-ionizing radiations (NIR) are associated with NGN. Social and societal aspects can also be critical: digital illiteracy and poverty can divide people, and lack of access to information, personal data collection and social surveillance could, if not properly handled, lead to a rejection of NGN.

The positive aspects of NGN are numerous and could offset the negative impacts. A few of them were presented in Section 1.4 and the rest of the book will provide more information on the positive aspects. Dematerialization, distance independence and process efficiency are the key words in summarizing the positive environmental aspects.

Telecom operators and manufacturers will have to determine their attitude toward such issues and set straight their corporate governance and operating standards. Dealing with customers and other stakeholders in a responsible manner will bring trust and confidence in corporations, thus allowing for a sustained development of NGN.

As the customer is the centre of attention in the operation model driven by customer needs, boundaries of activities should also be set through open dialogue with the stakeholders.

1.7 THE ORGANIZATION OF THE BOOK

This book introduces the world of NGN, starting from a distant perspective and narrowing to the core of the subject as the chapters progress. It is similar to discovering a new land: first it is observed from far as an overview, then in greater detail, revealing hills, valleys and forests.

This book is organized in the following 10 chapters:

1. Introduction
2. NGN Vision, Scenarios and Advances
3. NGN Requirements on Technology and Management
4. NGN Functional Architecture
5. NGN Operator, Provider, Customer and CTE
6. NGN Network and Service Evolution towards NGN
7. NGN Key Development Areas
8. NGN Standardizations
9. NGN Corporate Responsibility
10. Summary

In Chapter 1, the background to the birth of NGN is described, with the challenges traditional operators are facing and the transformation of operation model that they are undertaking. The objective of NGN is to improve the quality of life and to bring new life experiences.

In Chapter 2, a vision of NGN is proposed, which puts the end user (human) in the centre and mobilizes the NGN service capability for his or her benefit. Typical scenarios that could be realized thanks to the NGN are provided to inspire the imagination of reader. Compared with today's network, the advances brought by NGN are also mentioned.

In Chapter 3, derived from the NGN vision proposed in the previous chapter, the technology and management requirements for NGN are analysed and deduced. In an NGN environment, the customer-need-driven operation model sets very high requirements on management functions; fulfilling them will make the NGN come to life.

Chapter 4 covers the NGN functional architecture. Starting from the ITU-T NGN functional architecture, an NGN functional architecture with integrated management functions is proposed. Each functional component and the interfaces between functional components are explained.

Chapter 5 concerns the NGN operator, provider, customer and terminal equipment. It explains the NGN open interfaces (between functional components) that will create an extended landscape for operators, providers, customers and customer terminal equipment. Possible NGN operator, provider, customer and terminal equipment are speculated upon.

In Chapter 6, the major evolution steps to move from today's network toward NGN are highlighted – a roadmap for operators and other actors in this field.

Chapter 7 explains the key areas and key technologies to be developed in order to realize or implement NGN, considering today's situation.

Chapter 8 provides an overview of the current NGN-related standardization bodies, their activities, their achievements and their further plans.

Chapter 9 is on corporate responsibility, the attitude corporations may well be advised to follow in the face of the less positive aspects of NGN, such as energy consumption, resources depletion, waste management and social control. Depending on this attitude, trust could result and NGN may be well received and accepted.

Chapter 10 summarizes the whole book, highlighting the main issues of NGN that have been discussed and the contributions of this book.

2

NGN Vision, Scenarios and Advances

Since the invention of the telephone by Alexander G. Bell in 1876, the initial voice service, providing 'human to human voice communication', has evolved tremendously toward what is now called communication technology.

It has moved from short distance to long distance communications, from point-to-point to point-to-multipoint connections and further to multipoint-to-multipoint, from stationary to mobile user, from terrestrial only to terrestrial-to-air and further to terrestrial-to-sea communications.

The great appeal of this very basic service is that it broke the distance barrier and enabled a virtual voice communication environment.

Roughly a century later, information technologies were ripe enough to provide data communication services on a large scale for human-to-human and human-to-machine communications.

Since then, these IT data services have grown from the basic Internet services of email, FTP, Telnet and Web browsing towards the advanced communication services of VoIP, IPTV, instant messaging, multimedia conference call, etc. These services offer a broad range of applications that already enrich the life of the user.

It has become quite clear that the potential of services based on the Internet Protocol is unlimited considering its simplicity in connecting humans and objects, its flexibility in adding or removing media and its capability to build advanced services from simple service enablers.

Next Generation Networks: Perspectives and Potentials   Jingming Li Salina and Pascal Salina
© 2007 John Wiley & Sons, Ltd

Today, communication services are experiencing a third historic transformation, as CT and IT merge into ICT. ICT make networked IT services possible, where IT stands for services and computing and CT for data transportation.

Other factors behind this transformation reside in the huge progress in manufacturing technologies (automation, miniaturization), in increases in computing power and memory capacity, and in the development of software, better interfaces and batteries. The combination of these progresses has contributed to the spread of computers and peripherals among users.

In the following sections, we propose a vision of NGN derived from the logical consequences of the historical evolution of ICT. The purpose attributed to NGN is to serve human interests and desires and not the opposite. We have further developed some scenarios to provide details and hints to the reader to highlight the potentials and perspectives of NGN.

2.1 NEXT GENERATION NETWORKS: PERSPECTIVES AND POTENTIALS

Perspectives . . . :

Next Generation Networks will network any person, device and resource independently of distance, location and time, through integrated intelligent interfaces and with enriched media.

. . . . and potentials:

Next Generation Networks as a platform will offer ubiquitous connectivity and intelligent interfaces for human and machine communication as well as pervasive services access, bringing value to human life for its improvement and new experiences. This platform will also provide a 'playground' for everybody to create and deliver services to others.

We need to provide some detail on the potentials and the perspectives mentioned above:

- *Any person* includes children, adults and people with disabilities or technology-aversion.
- *Any device* includes sensors, terminals, machines and equipment.
- *Any resource* mean computers, databases, libraries, etc.
- *Independently of distance, location and time* covers space, the air, on land and underneath it, on the sea and submarine networking, on the move, nomadic or stationary.
- *Integrated intelligent interfaces* will break not only the technical and the language barriers, including written, spoken and sign languages, but also support media such as text, pictures, voice, video and other not as yet

commercialized media to convey information to all our senses, including smell, taste and touch besides vision and hearing.

The realization of this NGN vision will enable, among other possibilities:

- Virtual face-to-face communication to be established between humans far apart geographically and of different languages. Thanks to NGN, context-aware information will surround the persons and bring a feeling of near reality:
  - the fundamental barriers to human communication, including distance, language difference and disability will disappear;
  - new senses such as smell, taste and touch will complement the traditional use of vision and hearing;
  - context-aware information concerning the actual personal status can be added, including expected indications of location or language, but also more personal hints on his or her feeling and mood, based perhaps on voice analysis;
  - context-aware information concerning the local physical conditions can be added, covering local time, temperature, humidity, pressure and other weather and environmental conditions;
  - calling up a name, clicking on a picture, a symbol or a name, dialling a number, etc., will initiate the communication.
- Virtual environments to be created, joining together physically separated living spaces, moving vehicles, etc.; the persons involved can interact within the virtual environment, e.g. play chess on the same board. Thanks to NGN,
  - a virtual family environment will permit each family member to be included wherever he or she is, to share the family warmth, giving a feeling of being together, gaming together, watching television and talking together, partying together.
  - virtual environments will connect friends' circles, communities, corporate environments, distributed exhibitions stands, congresses participants, classrooms, etc.
  - new virtual worlds will be formed for the enjoyment of players or actors around the world (as in the game Second Life).
- Pervasive access to services, applications, content or information and seamless access even on the move across countries, the sea, land or in space.
- Anyone to create, develop and provide services, applications, content or information to a restricted group of users or to offer them in open access.

- Networked sensors to measure physical parameters locally and to analyse them globally. This can be used to:

  o predict or forecast natural disasters, earthquakes, tsunami, hurricanes, forest fires and other natural catastrophes by analysing information together via sensors embedded under the sea, in the air (in balloons) or under the ground;

  o generate context information that will be added to voice or video calls thanks to sensors included in a communication device;

  o collect health information through sensors placed in and around the body to report remotely any unusual status and to make further measurements and/or to release nano-medicines according to the instructions of a medical doctor;

  o collect home information with sensors located at various locations to monitor, report and act remotely;

  o collect vehicle information with sensors in vehicles, to report remotely any unusual status, to control driving direction and speed, and to enable automatic parking;

  o collect traffic information with sensors distributed along a road, to provide road congestion information and monitor flows of vehicles;

  o collect information from robots or unmanned vehicles exploring remote or hostile environments;

  o collect information on man-made or natural structures to monitor their deformation, state of strain and stability for security purposes.

- Networked cameras, a special type of sensors used to:

  o observe, with cameras installed on the ground or mounted on satellites, airplanes, balloons or ships, unusual phenomena or events such as unidentified flying objects, war or accidents, and report them directly and instantly;

  o activate a remote monitoring system in an airplane or any vehicle when an unusual status is observed, reported or forecasted;.

- Machine-to-machine communication to automate system managements in order to:

  o enable communication between ground vehicles for automatic traffic management, including driving at a safe distance, reporting accidents, broadcasting road traffic conditions or steering an automatic navigation system;

  o enable communication between airplanes for emergency help.

Figure 2.1 illustrates the future applications of NGN. The human-centred communication enables virtual realizations of space flying, congresses, exhibitions, teaching, home, offices, hospital and shops. It should be clear to the reader that most of these situations are feasible today or even to some extent

Figure 2.1   Applications of NGN

already a reality. Nevertheless, in most cases we observe dedicated solutions, self-contained systems, islands of information not networked due to technical differences and lack of standards. NGN brings solutions to these problems.

2.2   SOME POSSIBLE SCENARIOS

Today, the combination of a PC and Internet as a platform already offers a quite amazing breadth of services. Communication, information, education and enter-tainment are realizable; new technologies are appearing that can analyse one's patterns of behaviour and provide matched products and services; communities of interests form automatically, among other applications.

NGN will build upon this. Besides connecting computers and transmitting richer and context-aware information, it will also network new devices and machines to the service of people: it promises to bring additional value to human life by improving quality of life and by enabling new life experiences. Below are some possible scenarios.

2.2.1   Virtual Space Flight

 The virtual space flyer is seated in a real-sized spaceship simulator that communicates on-line with the spaceship and duplicates the same environment. The context information sent from the spaceship includes three-dimensional space views and physical parameters such as noise, temperature, pressure, humidity, acceleration and vibration, communications among the astronauts and their activities. In contrast, the transmission of a virtual home feeling (see Section 2.2.6) or simply the smell of home cooking could bring comfort to the astronauts! The virtual flyer can also talk with the astronauts on board or ask questions. All this will bring a feeling of reality in the simulator. NGN provides real-time communication, including context information. This differentiates this situation from standard simulations, usually reproducing a recorded situation (Figure 2.2).

Figure 2.2   Virtual space flight

2.2.2   Virtual International Congress

Participants in a congress do not need to physically attend the congress any longer. They can remain located around the world, at home, in offices, in local conference halls or attend while on the move.

Each congress participant uses his or her own language to deliver a speech and to join discussions. He or she receives speeches of others automatically translated into his or her native language. Each participant and his or her presentation are seen during the speech or called on request.

Papers are automatically translated, enlarged, read loud or transcribed for people with disabilities, and discussions are shared on a common board. Side meetings and personal contacts are possible through separate dedicated channels.

Compared with the congresses of today, requiring physical attendance, one will remotely attend even without knowing others' language, saving the travelling time and cost. It will greatly enhance attendance as the barriers are broken down (Figure 2.3).

Figure 2.3   Virtual international congress

2.2.3   Virtual Global Exhibition

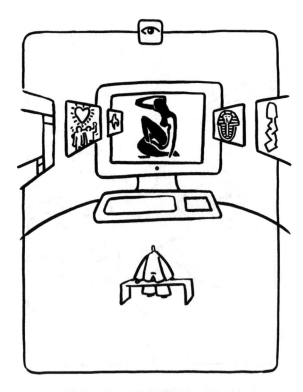 A potential customer expresses his or her needs on-line for a product or a service and thus launches thus a virtual call for tender. A virtual exhibition is then organized including all possible suppliers. Exhibitors who choose to bid stay at their own company premises from any corner of the world. Virtual visitors then choose the product of interest to them by clicking to gain information, including functional and non-functional characteristics like perhaps the weight, touch or smell. Demonstrations run remotely. Visitors can also require a person to have an on-line introduction to and discussion of the product.

Exhibitors organize panel discussions for the visitors in order to gain a comprehensive picture of different offers for the same type of product, not only in terms of price but also regarding the design philosophy, service, development roadmap and research.

Compared with today's conventional exhibition, visitors will obtain a comprehensive view of the technology or product concerned in very short time and again save the travelling time and costs. Exhibitors save the cost of setting up the exhibition stand and nevertheless bring sufficient information and demonstrations for their potential clients (Figure 2.4).

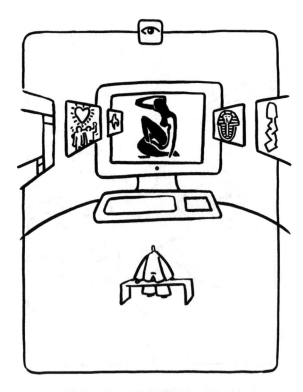

Figure 2.4   Virtual global exhibition

2.2.4 Virtual Classroom, e-Education and Experimental Laboratory

 Students attend lectures by the best professors in the field from their local classrooms in their home country. The students can ask questions of the professor or discuss topics with him or her in virtual face-to-face meetings. The professor also sees his or her students (and looks at their state of concentration, their interest and understanding level, how fast they take notes, etc.) and questions them.

The teacher uses resources such as libraries, three-dimensional demonstrations or multimedia located anywhere in order to demonstrate a point, enrich the lecture or simply keep the attention of students. In the experimental laboratory, the professor monitors the experiment of the student and provides personal advice on his or her skills development.

The languages used by the students and their professor can be different (Figure 2.5).

Figure 2.5   Virtual classroom, e-education and experimental laboratory

2.2.5   Virtual Corporate Environment

Global companies manage a very heterogeneous corporate environment of multiple locations, multiple languages and diverse staff and customers. The employee works in the corporate environment wherever he or she is, with access to intranet, working documents, projects, email and other corporate resources; scheduling and attending meetings; initiating on-line discussions with colleagues; involving people when needed; and sharing the same discussion board. Staff can access and re-create the virtual corporate environment when at home, in a hotel or on travel.

Productivity is increased and resource management streamlined as output is matched with input demand in quantity and quality independently of location (lean production in real time enterprises (Figure 2.6)).

Figure 2.6   Virtual corporate environment

2.2.6   Virtual Home

■ Feeling at home away from home will be possible, even in a spaceship (see Section 2.2.1) or after a real working day in a virtual corporate environment (see Section 2.2.5). The user enters his or her home environment to talk with the family members, to watch together the same television programme and discuss it, to play games with his or her children or to call grandma and grandpa with them. He or she participates in family parties and birthdays and recalls memories, getting the look, smell and taste of the family dinner while 'enjoying' a miserable sandwich away from home (Figure 2.7).

2.2.7   Virtual Hospital

✚ On-line medical doctors handle patients not at the hospital but wherever the patient is. In the event of a problem, networked sensors will send

Figure 2.7   Virtual home environment

a warning message to the responsible doctor. Depending on the seriousness of the case, he or she asks the patient to come over to the next hospital or even sends an ambulance for timely medical treatment. Otherwise, the patient is advised on what to do or medicines are provided *in-situ* through devices prepared in advance.

In emergency cases, such as accidents, shocks or strokes, since every second counts, the patient is first diagnosed remotely and is instructed what to do. An ambulance is then sent with the proper equipment to the location. Patients may not always be conscious: the proper handling of a patient in the ambulance with the help of on-line specialists or advanced warning from the ambulance to the hospital for the preparation of adequate treatment will save lives.

In difficult cases, medical doctors will consult their professional databases or specialists in order to set up a correct diagnosis, view similar cases and check for the most appropriate treatment at the closest available facility (Figure 2.8).

Figure 2.8   Virtual hospital

2.2.8   Virtual Store

 The consumer is shopping on-line. We have all already purchased books, DVDs, music and tickets on-line. These purchases required only one or perhaps two of our senses, vision and hearing. As we know, emotions or feelings enter for a great part into our decisions to purchase new items. This will be taken into account, as one will smell a new perfume before purchase or will touch and feel the textile of new summer clothes.

As the sizes of clothes are not yet standardized, it is still necessary to try them on: a virtual store offers the possibility to try new clothes in the comfort of home, checking the look from different angles, changing colours and patterns before sending an order to produce the clothes and receiving them at home a few hours later. A home tailor will scan our bodies for three-dimensional measurements and send them to the virtual store (Figure 2.9).

Figure 2.9   Virtual store

2.2.9 Global and Local Information Centres

 Networks of sensors collect global or local information via satellites, from airplanes, cities or mountains, or from under the sea. At the information centre, the relevant information from different sources is combined, processed and analysed. Thanks to a predictive model using artificial intelligence, natural disasters, human conflicts or, less dramatically, traffic jams, are forecast with a high confidence and early warnings delivered automatically to the relevant authorities in the expected language and format. Weather is forecast with a better accuracy and information forwarded to or accessed by the users.

People are concerned with people: virtual 'television studios' could be located at any place, in a private home, at the hotel or on the move. Eyewitnesses of an event, participants or would-be reporters can send their own content on-line to family, friends or anyone, or even add their own content to an on-going television programme (Figure 2.10).

Figure 2.10 Global and local information centres

2.2.10   Home Networks

 Aaah! Home networks – long dreamed of, the most popular item for prediction of the future! They always postulate a home filled with appliances designed to relieve women from the daily chores and bring comfort to men. A classic view of families enhanced with new technology! The appliances are networked; mobile devices can be added to the home network when at home and electronic devices integrated when switched on. The home gateway administers the entire home network and organizes traffic to the outside.

The home instruments include phones, PCs, televisions, electrical and electronic devices (refrigerator, monitoring cameras, sensors), our friend the home tailor (see Section 2.2.8) and heating or air conditioning systems.

The home gateway combines the function of a PBX and a portal. The gateway is called via a URL, connected to the personal phone of a family member. Family information is accessed at the click of a symbol, image or name. The home gateway can set access rights. The home network transmits information on the state of the house to the owner while he or she is away either on request or automatically (Figure 2.11).

Figure 2.11   Home network

2.2.11   Automatic Traffic and Car Driving
(Machine-to-machine Communication)

In automatic traffic, cars communicate constantly with other cars on the road, with sensors along the road and with the traffic data centre via NGN. The main purpose is to increase safety by maintaining safe distances and appropriate speed. Automatic warnings and even preventive action are issued if necessary to avoid accidents.

By communicating with the sensors along the road, driving at excessive and dangerous speeds is prevented. Car location is identified with better accuracy than with today's GPS or where GPS is not available, such as inside tunnels.

In constant contact with a traffic data centre, better route choice is possible for a driver, linked with personal preferences and contextual information: a car driver is informed on the intended driving road, either along the coast for leisure or reaching a destination in the shortest time considering the current and forecast road conditions (Figure 2.12).

Figure 2.12   Automatic vehicle

2.2.12   NGN Advances

NGN firstly creates a dedicated virtual communication environment where the usual barriers of distance, language or too rapid mobility (handover problem, Doppler effect) disappear. This virtual communication environment connects distant persons, as they will be talking face-to-face with communication content expressed for their five senses. A complete environment is virtually recreated on request or automatically; it joints separated spaces as if they were next to each other.

Secondly, NGN builds a universal platform where anyone can access existing services, applications, content or information and creates and delivers its own services, applications, content or information.

The fundamentals are laying on the communication with full human senses and surroundings, high intelligence, flexible connectivity, high quality, generic accessibility, generic mobility, open interface, high operation efficiency and high reliability and availability.

Communication with Full Human Senses and Surroundings

Today, the communication vectors are texts, voice, pictures and two-dimensional video, without communicating environment information. In an NGN environment, the content is transmitted for the full human senses, such as a feeling of dimension and space, smell, taste and touch, together with the context information.

The context information includes person-related context information such as mood, feeling, location and availability, and the environment-related context information (time of the day, temperature, pressure, noise, view, raining, etc.). Media can be added or removed during a communication.

The enriched communication in NGN can become undistinguishable from the real situation.

Virtual Living Environment Jointing Physically Separated Spaces

Today, there exist video call conference that shares the videos of each sepa-rated space. In an NGN environment, the virtual living environment can be formed from physically separated spaces; persons included in the virtual living environment can interactively do things together like living next to each other.

A separated space and person can be added or removed from a virtual living environment.

High Intelligence

Today's networking system does not enable communication in different languages. The user identification and service activation is intended more for machines than for human beings. In an NGN environment, the high network intelligence enables real-time communication in multiple languages, including sign language for speech- and hearing-impaired people. The language barrier has disappeared!

Furthermore the intelligence embedded in the network can identify end-users by biometric parameters from a face, a voice, an iris or eye, or a vein, and can allow the activation of a service with voice, sign language, hand-writing on-screen or a simple touch.

Flexible Connectivity

Today, networks are still dedicated to or designed for one type of service with specific connectivity. In the future, as connectivity and services will be separated, NGN will provide extremely flexible connectivity including:

- stateful or stateless connections;
- simplex or duplex connections;
- point-to-point, point-to-multipoint, multipoint-to-multipoint connections;
- symmetric or asymmetric connections
- person-to-person, person-to-device or device-to-device connections.

Connection(s) can be added or removed during a communication.

High Quality

Today, network quality has limitations due to bandwidth availability. Moreover, delay, jitter and packet loss cause distortion. NGN will provide extremely high networking quality for communication of content and context information that form a virtual communication environment similar to or undistinguishable from a real environment.

Generic Mobility

Today, service mobility is mostly limited to a single mobile communication system. NGN will provide generic mobility. This covers:

- global usability and reachability of a user for communication no matter where he or she is and if he or she is moving;
- communication seamlessness when a user is on the move across different access technologies, different networks and different countries, in space or moving across the sea.

Generic Accessibility

Today, each network is associated with its native service, typically GSM for mobile voice. NGN will enable access agnostic service accessibility. This describes the possibility to:

- access the same service/application/content/information via different access media, including optical fibre, twisted pair, coax cable, power line, wireless or satellite, and via different access technologies, including fixed, mobile and broadcasting;
- access a single service via the simultaneous use of multiple access technologies, for instance interactive television via the broadcasting access for receiving content and telecom access for responding;
- access the same service/application/content/information via different devices, including telecom devices, broadcasting devices or IP devices;
- access a single service via multiple devices such as a camera phone to send or a television to receive.

Open Interface

Today, there are only limited and primitive possibilities for a person to provide his or her own services, applications, content or information. The open interface of the NGN platform will enable every consumer to become a provider as well, to share his or her content with other audiences. When enjoying, for example, a television programme, the consumer's own content can be added to the programme and overlap with the original television programme under certain conditions.

High Operation Efficiency

Today, network resource usage is based on a 'first come, first served' basis. Only a few primitive measures are available for service priority, for instance voice has priority over data. In an NGN environment, the network resources assignment is prioritized according to a business strategy that includes the customer subscription, the business importance of the service or the business relationship with the provider.

Today, only a few quality of service (QoS) mechanisms are in place and cannot be activated in a consistent way, thus network resources are not always used in an optimized way. In an NGN environment, the QoS management mechanisms are placed along the end-to-end network chain in a consistent way to ensure that sufficient QoS is delivered with the service.

High Reliability and Availability

In order to manage the NGN, there are fundamental differences in operating the networks. For example, the self-organizing networks (SON) must be in place, especially the three following features:

- *auto-configuration and adaptation for systems and parameters* in order to reach (a) an optimized performance and (b) an optimized resource usage (requested service, demanded capacity and distribution);
- *a self-forming network,* when introducing any new nodes in the network;
- *a self-healing network,* when there is a networking problem.

Furthermore, proactive operations must also be in place: the network is constantly monitored and the performance trend analysed; measures are automatically activated before any degradation becomes noticeable.

3

NGN Requirements on Technology and Management

Until now, all the communication networks have been network technology-centred. This is because the network technology directly enables the services for end-users, thus generating the revenue for a network operator. The network management is just there to keep the technology working, playing a supporting role. Furthermore, the customer subscription has nothing to do with the running of the network. Technically all the customers are treated equally, i.e. first come, first served.

In an NGN environment, it is the customer subscription and the customer requested service that dictate the running of the network, that is, upon receiving a customer request for a service, the network will assign the service quality and service priority according to the customer subscription and the requested service, and the service priority is used for traffic engineering in case of network congestion. Under such a customer-centred approach, the management plays a role as important as the technology. It is a revolutionary step to transfer from a technology-centred approach to a customer-centred approach.

Considering the current situation, the realization of NGN will require tremendous technology advances, and much greater advances in management.

This chapter states the major requirements of NGN for both technology and management.

Next Generation Networks: Perspectives and Potentials   Jingming Li Salina and Pascal Salina
© 2007 John Wiley & Sons, Ltd

3.1   NGN REQUIREMENTS ON TECHNOLOGY

In the NGN environment, the technology facilitates the following:

- communication using the five human senses and surroundings;
- real-time communication across language barriers;
- virtual living environments joining physically separated spaces and enabling interaction of people;
- biometrics-enabled user identification;
- human-like service activation;
- on-demand connectivity;
- easy and standardized service creation;
- ICT-enabled customer terminal equipment.

3.1.1   Communication using the Five Human Senses and Surroundings

Today, multimedia communication includes only the hearing and visual human senses and no information about the person and the surrounding conditions. This is the preliminary approach to enabling human communication in a machine way due to the lack of essential technologies.

NGN communication will be able to include all five human senses and information about the person and the surrounding conditions as shown in Figure 3.1. This is an advanced approach to empowering the machine to enable communication between distant people that is indistinguishable, or almost so, from real face-to-face talking. This requires the technologies of:

- sensing, transmitting, transporting, receiving and re-generating communication content composed of the five human senses;
- sensing, transmitting, transporting, receiving and re-generating personal context information, like location, presence, mood and feelings;
- sensing, transmitting, transporting, receiving and re-generating on-going communication of information related to surrounding conditions, like local time (daylight, dark), temperature, pressure, humility, wind strength or vibrations;
- synchronizing all the information on communication content and context for real-time communication.

3.1.2   Real-time Communication across Language Barriers

Today, it is impossible for distant people to communicate in real-time in different languages. NGN will enable direct communication in different languages,

Figure 3.1 NGN communication with context information

including sign language, with embedded language intelligence. This requires the technology of:

- Recording speech accurately.
- Analysing and converting the recorded speech into a language-neutral data stream. Language expresses human thought. The same thought can be expressed in different languages. If a technology can catch the thought behind the language with a data stream, the language translation problem is then solved.
- Converting the data stream into the requested language.
- Extremely short processing times for the three steps above.

3.1.3 Virtual Living Environments

Today, so-called video conferencing provides only video sharing among different meeting spaces. No interaction is possible among the involved attendants from different meeting spaces, for example writing on the same board.

NGN will enable a virtual living environment to be created which joins physically separated spaces and enables interaction among the persons involved, e.g. playing chess together, writing on the same board while discussing, watching the same television programme and talking together, enabling grandparents to share in family warmth while they are in residential care.

The extended space can also be a room on the move, e.g. inside a car. This requires the technology of:

- large, high-definition screens and three-dimensional capable video,

  ○ SDTV (standard-definition television), ~2 Mbps per channel,
  ○ HDTV (high-definition television), ~10 Mbps per channel,
  ○ LSDI (large-screen digital imagery), ~40–160 Mbps per channel (ITU J.601);

- active screens which are capable of sensing a person's action and reproducing the action at a distance, e.g. to record the last chess position on the screen, convert it into a signal and regenerate it at the right position on the screen of another player – similar effects should be possible for multiplayer games;
- huge bandwidths, up to Gbps connectivity of high quality;
- flexible multipoint-to-multipoint connectivity, the connectivity being set up according to the capability of the end-user device;
- seamless mobility support, i.e. when a train is moving at high speed, such as 500 kmph, the connection remains constant and without noticeable quality degradation in terms of bandwidth, delay, jitter or packet loss.

3.1.4   User Identification using Biometrics

Today, a user is identified by typing a code or password, a machine-like approach. NGN will enable easier, faster and more accurate means of human-like user identification, i.e. a three-dimensional face, voice, iris or eye, a vein under a finger. This requires the technology of:

- recording a person's voice with identifiable characteristics;
- taking three-dimensional pictures of a person's face with identifiable characteristics;
- taking a picture of the pattern of a person's veins under the skin, which is as identifiable as a fingerprint;
- comparing and analysing data with the reference, with noise introduced by voice distortion due to sickness or aging, face deformation due to temporary skin problems or aging, etc.

3.1.5   Human-like Service Activation

Today, service activation is done in a machine-like way, e.g. by clicking or typing words. NGN will enable more human-like service activation, e.g. speaking a name to initiate a call, sketching on the phone screen to send a personal SMS or a personal drawing, speaking or writing a key word or a group of key words to activate a sophisticated service or to call up information. This requires the technology of

- speech recognition with ~100 % accuracy;
- grabbing handwriting or hand-drawing from the screen of a device;
- semantic search.

3.1.6   On-demand End-to-End Connectivity

Today, a physical communication network is capable of a specific type of connectivity required by the network-associated communication service, e.g. GSM's narrow-band duplex symmetric point-to-point wireless connectivity or DVB's broadband simplex point-to-multipoint connectivity. In an NGN environment, the transport network is separated from services and targeted at making use of all the available physical media to provide the end-to-end connectivity according to the service requirements. This requires the on-demand formation of end-to-end connectivity with flexibility in connection type, quality and security as shown in Figure 3.2, i.e. NGN is:

- capable of setting up point-to-point connections, point-to-multipoint connections and multipoint-to-multipoint connections; for point-to-multipoint and multipoint-to-multipoint connections, an individual connection can be added or removed during an on-going communication;
- capable of simplex or duplex connection, where
  - ○ simplex connection is for communication in single direction,
  - ○ duplex connection is for communication in both directions;
- capable of symmetric or asymmetric duplex connection, where
  - ○ symmetric connection provides the same bandwidth in both communication directions,
  - ○ asymmetric connection provides different bandwidths in both communication directions;
- capable of end-to-end connectivity with fine quality granularity including bandwidth, delay, jitter and packet error rate from very low performance to extremely high performance, e.g.,

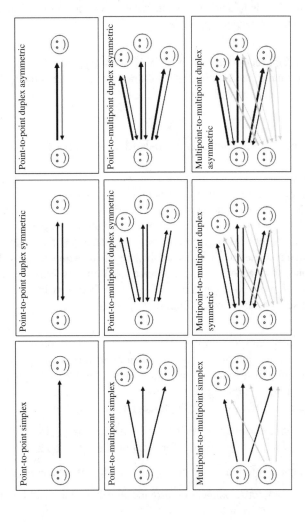

Figure 3.2 NGN flexible connectivity

- o very low performance connection for GSM SMS, which requires even 3 dB less signal strength than GSM voice,
- o extremely high performing connection for Grid computing to process astronomic data at different places;

- capable of providing quality-guaranteed, secured and reliable end-to-end connectivity across heterogeneous networks over technologies, operators, countries and continents via standardized interfaces at each part of the end-to-end connectivity;
- capable of using simultaneously multiple access technologies for a single service, e.g., interactive mobile television:

- o receiving television programme via DMB satellite or terrestrial access,
- o responding for voting via fixed or mobile access;

- capable of simultaneously using multiple devices for a single service, e.g. receiving a television programme at a television set or voting via fixed or mobile phones.

3.1.7 Easy and Standardized Service Creation

Today, most services are associated with the networks. The associated service lasts as long as the network exists. Value added services (VAS) are added in a proprietary way and in a silos manner.

In NGN, service is no longer lifetime – a service comes and goes, but the carrier network remains. Therefore, service creation has to be standardized and hidden from the complexity of a network, i.e. one does not need to understand the network to create services, applications, content or information and to deliver them via a network or networks. This requires the technology of:

- Open APIs (application programming interfaces) between the application layer and the service stratum, as shown in Figure 3.3. The open APIs provide a standardized way for an application, for example, to call a basic network service to set-up a call, to send an SMS or to fetch the current location of a person, and the APIs are programmed in Java, and are therefore easy to use. Therefore, open APIs enable the standardized and easy creation of advanced services or applications that are independent of the underlying network; it thus enables network-agnostic service access.
- Rich network service enablers for building up the most advanced services and applications. This depends on the network operator to provide basic network services or network service enablers. The network service enablers can be:

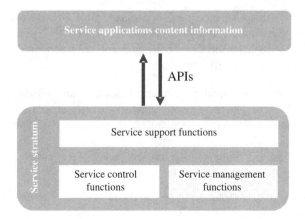

Figure 3.3   NGN open interface and enablers for service creation

o   for call set-up;
o   for session set-up;
o   for location/presence/availability information;
o   for charging.

3.1.8   Flexible Terminal Equipment

Today, customer equipment remains at a preliminary stage – telecom terminals are built for telecom services and IT terminals are built for IT service. It is not only a problem of duplicating service, but also causing confusion and inconvenience for the end-user.

The NGN terminal equipment is the first or last element of NGN, interfacing with the end-user to initiate and to receive ICT services, and eventually providing the ICT service. It can also act as a gateway for other device(s) to be connected to the NGN.

Besides human-like user identification and service activation, such ICT-enabled customer equipment requires the technology:

• As a home terminal device –

   o   for traditional services, e.g. voice, an end-user is able to select the quality level in a simple way, e.g. excellent or good, where the technology behind is hidden;
   o   an end user can choose the communication media, e.g. voice-only or video and voice;

- an end user can upload a self-designed interface with a welcome message or family album or video for recent family events with defined accessibility; this interface will appear when being called by others;
- an end user can access any service, application, content and information; the network will automatically make the reconfiguration and upgrade or download the software for the called service or application.

- As a home gateway –

  - it is capable of multitechnology connectivity (mobile, wireless, infrared, ZigBee, etc.);
  - it is capable of making an ad-hoc connection for a device under its coverage to the NGN, e.g. when the mobile is moving into its coverage, the call will be routed via it instead of consuming mobile network resources;
  - it is capable of organizing the connections among the devices under its coverage;
  - it is capable of using multiple devices for a single multimedia service, e.g. television for showing video or computer for gaming.

- As a mobile device –

  - it is capable of reporting terminal capability to the network;
  - it is capable of providing information about terminal location (e.g. A-GPS) or moving speed on request;
  - it is capable of reporting the current available access technology to the network;
  - it is capable of measuring the QoS received at the terminal and reporting to the network, e.g. downlink MOS (mean opinion score) or downlink throughput;
  - it is capable of detecting neighbouring device(s), and communicating directly with them via short-range radio technologies, e.g. UWB, Bluetooth or WLAN;
  - it is capable of detecting neighbouring device(s), to form an ad-hoc network or to join a local network, e.g. CAN, BAN, PAN, HAN, therefore enabling the telecom service to be available for other devices within a local network.

- As an enterprise gateway or IP PBX –

  - it is capable of offering all current PBX functions, including short number dialling and hunting;
  - it is capable of providing the company home page with advanced capabilities to present company information with webinar, on-line exhibition, on-line discussion, etc.;

3.2   NGN REQUIREMENTS ON MANAGEMENT

As mentioned at the beginning of this chapter, NGN requires advanced customer-centred management along the end-to-end service delivery chain as shown in Figure 3.4 that facilitates:

- customer management;
- third-party provider management;
- service and service delivery management;
- network and network performance management;
- security management;
- information management;
- customer equipment management.

3.2.1   Customer Management

Today, customer management is no more than subscription, provisioning, billing, etc. In an NGN environment, under the customer-centred approach, customer management becomes a central task of any operator and is required to be:

- capable of clustering the customer profile with the service favourability to be able to pick up the right customer group for a campaign for a new service, or for more customer-oriented service design;

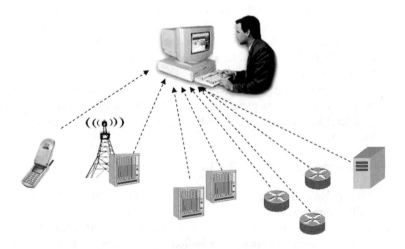

Figure 3.4   NGN service delivery platform

- capable of predicting the customer loyalty or the churn rate by analysing historic data, including age, subscription type and duration as well as the customer's service usage and payment record, in order to take early measures to prevent the loss of the customer;
- capable of changing the customer subscription and service profile easily on-line and making it instantly effective;
- capable of monitoring the SLA satisfaction level and providing it to the customer on request;
- capable of providing the customer with the possibility to check the account at any time via the Internet or other means;
- capable of flexible and comprehensive billing, including:

  o billing in real-time including prepaid or post-paid,
  o billing business customers with special tariff,
  o subscription class-based billing,
  o service type and usage-based billing, e.g. accumulated service usage and combined service usage,
  o calling and called party relationship-based billing,
  o billing for using the service on specific dates, e.g. birthday or for promotional campaigns,
  o SLA-based billing, i.e. when the SLA(s) is not fulfilled at the end of period, a penalty has to be paid to customer,
  o QoS-based charging, i.e. when service is not delivered with sufficient quality, it will be charged partly or not charged at all,

- capable of selecting targeted customers for a promotional campaign;
- capable of customer self-management on-line

  o for changing subscription or subscription class
  o for changing service provisioning or service profile
  o for defining closed user groups to access personalized or specific information like location, presence, availability, etc.
  o for checking account status,
  o for checking QoS delivered with the service,
  o for checking the SLA satisfaction level,
  o for self-administration of PBX-like services, e.g. adding and removing employees, short number plans or black and white lists for out-going calls.

3.2.2   Third-party Service Provider Management

Today, there are already third-party providers of service, applications, content and information. However there is no systematic management of them. In an NGN environment, third-party providers will be contributors of service, applications, content and information for the end-user. Actually, each end user can be a third-party provider. This requires the management of:

- registration and resignation of third-party provider;
- easy discovery and usage of available network service enablers;
- billing mechanism for the usage of network service enablers;
- monitoring of SLA satisfaction from both sides.

3.2.3   Service and Service Delivery Management

Today, a network is designed and built to deliver a service, i.e. the specific service is only available where the network goes. For other services added later, there is no service delivery management.

NGN requires pervasive service access, i.e. the same service is available via different underlying networks, possibly with different service qualities. This requires the service delivery to be independent of the underlying network, efficient and reliable, i.e.

- capable of adding, maintaining and removing services available to the end users;
- capable of delivering services/applications/content/information to any end-user from any provider according to the SLA;
- capable of setting up the service delivery requirements in terms of QoS, security and reliability according to the customer subscription, the service quality requirement and the business importance of the service, e.g. the business relationship with the service provider;
- capable of acquiring information about the terminal service capability, when necessary, to enable remote configuration or re-configuration over the air (OTA);
- capable of adapting the delivered services/applications/content/information in the right format and the right media according to the capabilities of the customer terminal equipment;
- capable of delivering rich media services/applications/content/information to the right device(s), in the case where a single service simultaneously involves multiple devices, e.g. video to television, smell to PC, talk via phone;
- capable of delivering context information with the communication content;
- capable of delivering critical information like natural disaster forecasts to the relevant user;
- capable of monitoring the delivered service quality to the end-user;

○ the load on the servers,
○ the performance of the servers,
○ the performance of end-to-end connectivity,
○ the performance of the terminal.

- capable of analysing the collected service quality data and platform performance (via monitoring) to predict the potential service delivery problem and
activate preventive measures;
- capable of detecting a problem occurring, analysing the reason(s) for it
and activating recovery measures; analysing the impact on services and
customers and providing organized information for management, operating
personnel and customer care;
- capable of producing information for QoS-based charging.

3.2.4   Network and Network Performance Management

The quality delivered with the service depends on many factors, however
network performance is the most decisive among them. The network performance refers to the connection quality between terminal and terminal or
terminal and machine. The NGN requires high-quality service delivery and
thus naturally demands a high quality of network performance. It needs
to be:

- capable of adding, maintaining and removing the new/old access network,
the transmission network and the transport network without affecting the
delivery of service/application/content/information;
- capable of managing multiaccess networks and multitransport networks
which belong to different operators, including

○ registering and de-registering an access network,
○ monitoring the availability, load and performance of all the access
networks,
○ registering and de-registering a transport network,
○ monitoring the availability, load and performance of all the access
networks;

- capable of support interoperability and interworking with NGN-compliant
networks, e.g. 3GPP CNs (at different phases), 3GPP2 CNs, external IP
world and legacy systems;
- capable of setting up end(s)-to-end(s) connectivity with the optimum access
and transport networks according to the service requirement provided by
service management, including

- o choosing the access network according to the user status (stationary or on-the-move, moving in the air or moving under the water), terminal capability, performance requirement and load,
- o choosing an adequate transport network according to the access network, interworked external network, performance requirement and load,
- o communicating the requirement to the interworked external network;

- capable of maintaining the connectivity for an end-user moving at high speed across different radio access technologies and networks (inter- and intrasystem handover, roaming) without noticeable performance degradation of the connectivity;
- capable of monitoring the performance of end-to-end connectivity, and activating the necessary measures when a potential or actual problem is dictated;
- capable of providing consistent control mechanisms across access, transmission and transport, which are embedded in the network node to realize traffic prioritizing for end-to-end connectivity;
- capable of managing heterogeneous networks;
- capable of interworking with interconnected networks across operators, across countries or across continents;
- capable of monitoring the interconnected network performance (under the restrictions of the SLA);
- capable of analysing the collected past and current data to predict potential problems in the network, and activating preventive measures;
- capable of detecting problems when they occur, analysing the reason(s) for them and activating recovery measures; analysing impacts on services and customers and providing essential information to management, operating personnel and customer care;
- capable of acquiring information about the terminal connectivity capability (capable access technologies), access technology availability and network load;
- capable of predicting network capacity according to traffic load monitoring and analysis.
- capable of tuning the network performance according to need.

3.2.5   Network Security Management

The network must be:

- capable of recording and identifying security problems;
- capable of designing new security measures when a new case is encountered;
- capable of detailed data inspection to stop traffic from certain providers or certain types of services;

3.2.6 Device Management

The network must be:

- capable of configuring/re-configuring device OTA or via other medium for operating systems (OS), software (SW) and application SW;
- capable of monitoring device performance and evaluating the terminal's merits (in a statistical sense);
- capable of detecting a device with a problem, and informing the user of a device change;
- capable of providing terminal capability for the correct media and format delivery of services/applications/content/information;
- capable of disabling stolen devices.

3.2.7 Information Management

The network must be:

- capable of providing centralized static information about the customer, customer subscription, SLA, terminal type, terminal capability and re-configurability, network nodes, server, network topology and system topology;
- capable of providing centralized dynamic information for customer location and availability, current SLA satisfaction level, network resource availability, service resource availability and current problems;
- capable of processing and analysing collected data from a broad base (customer, service, terminal, network, service platform, traffic, load) to provide historic statistics, discover the co-relationship and therefore predict the tendency for network performance, delivered service quality, etc.
- capable of providing traffic information detailed to the user or type of service.

4

NGN Functional Architecture

Today's network was built originally for one dominant service for the lifetime of that network; therefore there is no real concept of architecture. The services or applications added later were implemented in a proprietary way in the silos manner, i.e. each service or application has it own billing, management, etc. Such an approach not only heavily multiplies the same functionality, but it has also brought the system to such a messy state that:

- a provider does not know the reason if a service is not running, so there is no way to fix the problem;
- to add a new service is a game of plug and pray.

NGN will come with a well-defined architecture. It is advanced in that it promises:

- the simplicity and flexibility to add/maintain/remove service, application, content and information;
- the easy creation of advanced service/application/content/information.

4.1 THE ITU NGN FUNCTIONAL ARCHITECTURE

The NGN functional architecture has been agreed among the different standardization bodies. Based on the ITU-T NGN Functional Architecture [1,2], with further simplification and complementarity, the NGN functional architecture is abstracted in Figure 4.1.

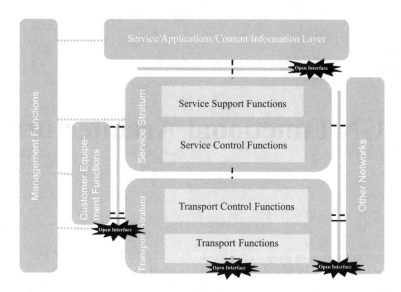

Figure 4.1 ITU NGN functional architecture. Reproduced by kind permission of ITU

The NGN is composed of the domains of *service stratum, transport stratum, customer terminal equipment functions, service/application/content/ information layer, management functions* and *other networks*. The thick grey lines mark the open interfaces between domains. The dashed grey lines mark the management data paths. The thin black lines mark the user data paths. The dashed black lines mark the signalling data paths.

The remarkable aspects of this NGN architecture are:

- The open interface between the *service stratum* and *service/application/ content/information layer*, which enables

  o a standardized application creation/execution environment;
  o easy service creation using application programming interfaces (APIs), which can be called with high-level programming languages, e.g. Java;
  o various kinds of providers for service/application/content/information, including internal or third-party, trusted or non-trusted providers.

- The separation of the *service stratum* and *transport stratum*, which enables:

  o the flexibility to add/maintain/remove service/application/content/ information without any impact on the transport layer;
  o the flexibility to add/maintain/remove transport technologies without any impact on the access to service/application/content/information;

- o optimized usage of multiple access and core transport technologies to form end-to-end connectivity across multiple terminals, different access technologies and different core transport technologies.

- The separation of the *service support layer* and *service control layer* within the *service stratum*, which enables:

  - o the service creation/execution functions to be separated from service delivery functions;

- the separation of the *transport control layer* and the *transport layer* within the *transport stratum*, which enables:

  - o the end-to-end connectivity to be set up and maintained according to the requirements given by the *service stratum*;
  - o optimization of the usage of the available network resource.

- The open interface between the *access transport* and the *core transport*, which enables:

  - o any available access network to be combined with any available core transport network to form the end-to-end connectivity;
  - o new access technology to be added; old ones can be maintained or removed;
  - o new core transport technology to be added; old ones can be maintained or removed.

- The open interface towards other networks, which enables:

  - o interconnectivity across heterogeneous networks based on different technologies, from different operators or from different countries;
  - o interconnectivity with understandable QoS requirements;
  - o interconnectivity with understandable security requirements.

- The open interface between the network and *customer equipment*, which enables:

  - o the same services to be available from different types of terminals, e.g. fixed or mobile phone;
  - o the same services to be available via different types of local area network (LAN) gateways, including PBX for Enterprise Network, a set-top box for a home network or an access point for the LAN, e.g. car, bus or plane area network.

- The NGN management covers all the NGN domains.

4.2 THE PROPOSED NGN FUNCTIONAL ARCHITECTURE

Compared with the original ITU NGN functional architecture, the proposed NGN functional architecture has great detail in the management part, which is key to realizing customer-centred operation.

Today, the management functions are carried by OSS and BSS, in which:

- the OSS is responsible for everything related to technologies, including network configuration and performance management;
- furthermore, performance management is done in a passive way, i.e. action is taken only after a problem has appeared;
- the BSS is responsible for everything related to customers, including customer subscription, service provisioning and billing.

In principle, there is no real-time direct communication between OSS and BSS. Technically all the customers are treated equally following a first come, first served approach.

There do exist a few primitive service prioritization or discrimination mechanisms, e.g. voice has higher priority than the data. However such mechanisms are hardly configured in the network, and there is no on-line if – then action.

The NGN will require the network management to enable the customer and service-dictated OSS, i.e. the network runs according to the actual business policy and a particular customer or service can be given higher priority for the network resource assignment, which also means the service quality. Such business policy is the implementation of business strategy of an NGN operator, which assigns network resources according to predefined criteria based on:

- who the customer is, what subscription he or she has, what the current satisfaction level of his or her SLA is;
- which service is requested, what the business important of the service is, what the QoS requirement of the service is, what the business relationship with the service provider is;
- what kind of terminal the customer is using in terms of multimedia capability, multiradio access technology capability, etc.;
- what kind of access network resource is available;
- what kind of core network resource is available.

That is to say, in the NGN environment, not only do the OSS and BSS have to talk to each other, but also the BSS is in fact dictating the network operation.

To realize such NGN management, we suggest splitting the management functions in the ITU NGN functional architecture into the *network management*

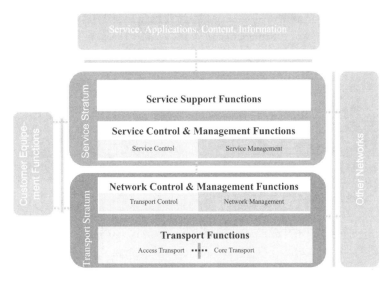

Figure 4.2 The suggested NGN functional architecture with integrated management functions.

part and the *service management* part, and integrating the *network management* part into the *transport stratum*, and the *service management* part into the *service stratum*. The suggested functional architecture is illustrated in Figure 4.2; the functionality of each element is explained in the following sub-sections.

4.2.1 Transport Stratum

The *transport stratum* is responsible for providing end-to-end connectivity according to the service requirements, the terminal capability and status and the network and resource availability. The *transport stratum* is further divided into the *transport control layer* and the *transport layer*.

Transport Control and Management Functions

The *transport control and management functions* are responsible for setting-up/ maintaining/terminating the end-to-end connection according to the service requirement, the terminal capability and status and the network and resource availability, i.e. telling the terminal which access network to connect to, the access network which core network to connect to and the terminal, access network and core network what type, quality and security of service to provide.

To fulfil this task, the *transport control and management functions* rely on the *transport control* and the *network management*. The two groups of functions

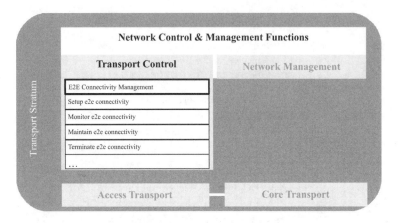

Figure 4.3 NGN *transport stratum – transport control* (e2e, end-to-end)

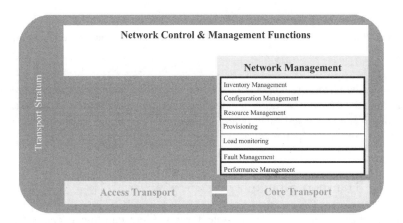

Figure 4.4 NGN *transport stratum – network management*

are interactive. The *transport control* is shown in Figure 4.3 and the *network management* in Figure 4.4.

Transport Control

The *transport control* manages the end-to-end connectivity in the following way:

- setting-up the end-to-end connectivity based on the service requirement, the terminal capability and status and the network and resource availability, where

- o the service requirement is provided by the *service management functions* of the *service stratum*, which concerns the priority, connection type, QoS parameters and security. The priority is set according to the business strategy, based on the business importance of the service, the business relationship with the provider, the customer subscription class and the current SLA satisfaction level of the customer. The connection type is simplex or duplex, point-to-point or point-to-multipoint or multipoint-to-multipoint. The QoS parameters are bandwidth, delay, jitter and packet loss rate. Security concerns confidentiality, integrity, etc.
- o the terminal capability and status and the network resource availability are provided by the *network management functions* at the *transport stratum*. The terminal capability and status concern the access technologies, display screen, and speed which are known at the network attachment and through the location update. The network and resource availability concerns the available access and core networks and the network load along the end-to-end connection chain, which is known via the constant monitoring of the access and core networks.

- maintaining the end-to-end connectivity,

  - o when any problem is monitored, the *transport control functions* are informed, and a re-set-up of end-to-end connectivity is initiated;
  - o when the network resource is requested by another end-to-end connectivity with higher priority, a re-set-up of end-to-end connectivity is initiated;

- terminating the end-to-end connectivity when

  - o the service is finished normally;
  - o the network resource has to be given to another request with higher priority.

- monitoring the end-to-end connectivity performance,

  - o after the set-up of an end-to-end connectivity, its performance is constantly monitored.

- managing the mobility of inter-access networks,

  - o when a terminal is moving across access networks from different access technologies or different operators.

Network Management

The *network management* is for:

- Inventory management – automatic and active inventory management by:

  - o defining a capacity threshold – when the traffic load is constant over the threshold, an automatic network resource provisioning is initiated;

- o using Agent technology to send inventory inquiries to all the network nodes, to get the actual topology of network, the HW, SW and IP address of the nodes, etc.
- Configuration management – centralized and remote configuration.
- Resource management – to monitor constantly the availability and the load of networks.
- Performance management,

  - o to monitor constantly the performance of every involved access network and core network;
  - o to analyse the performance tendency, predict the potential performance problem and activate measures in advance;
  - o to try to recover the monitored problem by activating the pre-defined measures, and inform the relevant teams with relevant information.

Transport Functions

The *transport functions* are composed of access and core transport networks, as shown in Figure 4.5. They carry the traffic end-to-end with the pipe ordered and controlled by the *transport control layer*. They fulfil their task by:

- forming the required pipe to move packets with the right access, transmission and transport network;
- engineering the traffic with embedded QoS mechanisms to transport packet according to priority;
- engineering the traffic with embedded inspection to filter the traffic;
- counting the transported packet for billing.

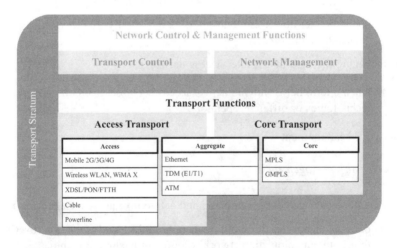

Figure 4.5 *Transport functions* with *access* and *core transport*

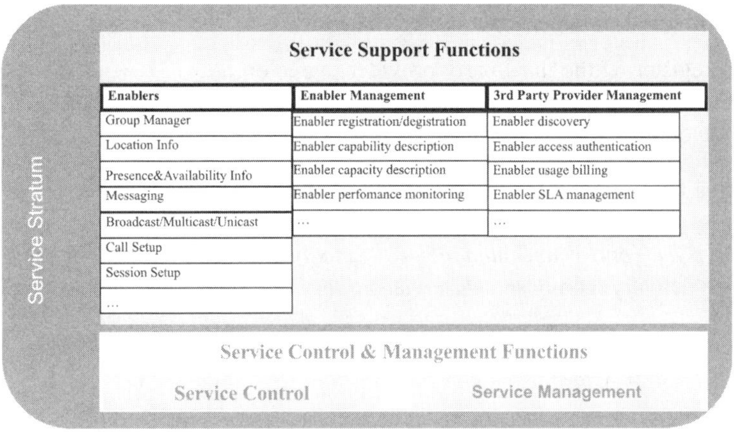

Figure 4.6 NGN *service support functions* in *service stratum*

4.2.2 Service Stratum

The *service stratum* is responsible for enabling the creation and the delivery of service, application, content and information. It is further divided into the *service control and management functions* and the *service support functions*.

Service Support Functions

The *service support functions* are responsible for providing the enablers for the creation of service, application, content and information to the internal or third-party provider, and for managing the third-party providers.

The *service support functions*, as shown in Figure 4.6, are composed of the enablers, the enabler management and the third-party provider management. An enabler can be added, maintained or removed according to need. The important point is that the enablers can be called with open APIs, where 'open' means 'standardized'.

- the security level is based on the requirement of service as well.
- the reliability is based on the requirement of service as well.
- Network service enabler management handles:

  - enablers registering;
  - enabler discovery;
  - monitoring the usage and performance of enablers;
  - provisioning more capacity for an enabler when it is being constantly overloaded.

- Third-party provider management handles:

  o registering the third-party provider (agree on SLA);
  o authenticating the third-party provider when using the enablers;
  o monitoring the performance of the third-party server.

Service Control and Management Functions

The *service control and management functions* are responsible for the NGN service delivery, composed of *service control* and the *service management*.

Service Control

The *service control* is responsible for delivering the NGN services and is composed of the service components of, for example, IMS (SIP-based multimedia services), PSTN/ISDN emulation, streaming and other multimedia services, as shown in Figure 4.7.

- The *IMS service component* is for SIP-based IP multimedia services, e.g. push-to-talk [3,4].
- The *PSTN/ISDN simulation component* utilizes the IMS capabilities to provide PSTN/ISDN-like services for advanced terminals such as IP-Phones or for terminal adaptations connected to legacy terminals, using session control over IP interfaces and infrastructure.
- The *PSTN/ISDN emulation component* supports legacy terminals to connect to the NGN; the user should have identical experience as provided by the legacy PSTN/ISDN services.

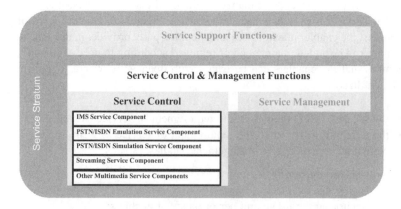

Figure 4.7   NGN *service control* within the *service stratum*

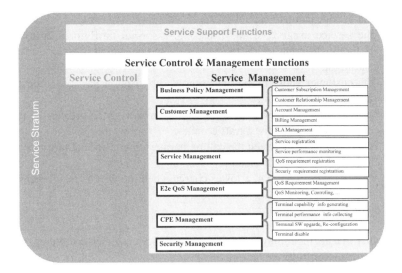

Figure 4.8   The NGN *service management* within the *service stratum*

A service component can be added, upgraded or removed.

Service Management

As shown in Figure 4.8, the *service management* assigns the network resource according to the priority level of the requested service, to generate the requirement on the end-to-end connectivity, to ensure the service is delivered to the end-user according to the business policy, and to bill the customer fairly for the use of the service.

The functions can be divided into *business policy management, customer management, service management, end-to-end QoS management, customer terminal equipment management* and *security management.*

Business Policy Management

Business policy is the realization of the business strategy of an NGN operator. The business policy is a set of rules:

• To define the service priority level according to the business importance of the requested service, the customer subscription, the business relationship with the service provider and the current SLA satisfaction level of the customer. The service priority level is provided to the *transport control* to assign the network resources for the end-to-end connectivity.
• To define the billing method based on:

- o the service campaigns;
- o the service usage date and time, e.g. a special tariff for a customer's birthday, different tariffs for weekdays and weekends, daytime, evening and night;
- o the customer subscription, e.g. a gold subscription has lowest cost per call or per traffic unit;
- o the accumulated service usage of a customer, e.g. the greater the usage the lower the price for a traffic unit;
- o the relationship between the calling partners, e.g. a special tariff for calls between a wife and husband.
- o QoS-based billing, e.g. a lower charge when the QoS is insufficient.

Customer Management

Customer management plays a key role in an NGN customer-centred operation environment, covering:

- customer subscription;
- customer service provisioning;
- customer self-care (on-line subscription, service provisioning, etc.);
- customer account;
- customer billing;
- customer data (profile, identity, number, address, language, service profile, service usage, service customization, payment moral, location, presence, availability, etc.);
- customer churn rate analysed by constantly monitoring the customer service usage pattern and the tendency to activate a measure necessary to keep those possible churn customers;
- customer's SLA satisfaction level monitoring, to pay a penalty when the SLA is broken, and to activate special measures when the critical level has been reached, e.g. changing the service priority level for the concerned customer.

Service Management

Service management involves:

- registering service, application, content and information available for end users;
- recording the QoS requirement for each registered service including band-width, delay, jitter and packet error rate;
  - o at the request of each service, the QoS requirement will be provided to the transport control to set up adequate end-to-end connectivity;

- recording the security requirement for each registered service, including data confidentiality and data integrity;
- monitoring the availability and the performance of each registered service, application, content and information.

Quality of Service Management

Quality of service management involves:

- monitoring the service quality delivered to the customer – when the delivered QoS is not sufficient, information is generated and sent to the billing system;
- predicting the QoS tendency by analysing the collected data from the transport performance and from the server performance.

Customer Terminal Equipment Management

Customer terminal equipment management involves:

- providing the customer terminal equipment (CTE) capability for the requested service according to the terminal type, SW version and configuration;
- reconfiguring the CTE for the requested service;
- upgrading the SW or client for the requested service.

Security Management

- Service access control – only subscribed and authorized customers including end-user and third-party applications can access services, applications, content and information.
- Confidentiality and integrity – to guarantee the safety of data stored and in transit, preventing data sniffing, spoofing, modification and so on.
- Security policy – to consider the trade-off between security and performance; to apply only the necessary security measures to an application according to the trust level of the application provider (non-trusted, trusted, internal), to the security sensitivity of the application (confidentiality, restriction, openness) and to the financial aspect (payment for the data security).
- Accountability for attempted attack and resource – to record any security-related event and trace the attempted attack and the crackers.

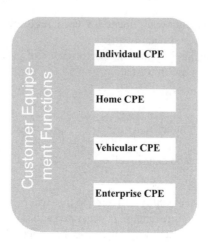

Figure 4.9 Customer terminal *equipment functions*. Reproduced by kind permission of ITU

4.2.3   Service/Application/Content/Information Layer

The Service/application/content/information layer is composed of servers, which could be from the network operator or the service provider itself, or from third-party providers. The important point is that the servers can utilize the open APIs to discover and use the available enablers, and to create and execute their own service/application/content/information.

For third-party providers, the trust is ensured by the SLA between the operator and the third-party provider.

4.2.4   Customer Terminal Equipment Functions

Depending on the type of customer, the NGN customer terminal equipment can be an end-user terminal device, a gateway for a corporate network, a gateway for a home network or a gateway for a vehicle LAN, as shown in Figure 4.9.

NGN customer terminal equipment will be ICT-enabled, i.e. services are IT-based and connectivity CT-based, but with the capability to simulate or emulate the traditional telecom communication services.

For individuals, the customer terminal equipment comprises the end user devices or home devices, for example:

- legacy terminals, e.g. mobile phone, PSTN telephone, Fax machine, text phones (working with a home gateway);
- SIP phones;

- soft-phones such as a programmed phone on a PC;
- IP phones with text capabilities;
- set-top boxes;
- multimedia terminals;
- PCs;
- user equipment with an intrinsic capability to support a simple service set;
- user equipment that can support a programmable service set.

For an enterprise or organization, the customer terminal equipment is a gateway which is capable of providing the services being provided today by integrated PBX and company webs separately, for example:

- short number dialling;
- hunting list;
- webinar;
- on-line exhibition.

For home networks, the customer terminal equipment comprises the home gateway and home devices which are capable of providing the services being provided today by telephone and home webs separately, for example:

- short number dialling for each family member;
- pictures or video about recent family event;
- family news.

For vehicles, the customer terminal equipment comprises the vehicular gateway and vehicular devices which are capable of providing the services being provided today by telephone and home webs separately, for example:

- sightseeing with the introduction of location based information and multiple language choice;
- multimedia broadcasting about a special event.

4.2.5 Other Networks

- Other NGN networks with multiple NGN administrative domains.
- Other non-NGN networks, including:
  - other IP networks and by implication any IP-based network which complies with the NGN interconnection protocol suite;
  - PSTN/ISDN by means of interworking functions that are implemented within NGN;
  - PLMN networks;

- broadcasting networks;
- enterprise networks;
- cable networks;
- the Internet.

REFERENCES

[1] ITU GSI-NGN Release 1 part I.
[2] ITU GSI-NGN Release 1 part II.
[3] Gonzalo Camartillo *et al. The 3G IP Multimedia Subsystem*. Wiley: Chichester, 2004.
[4] Miikka Poikselkä *et al. The IMS – IP Multimedia Concepts and Services in the Mobile Domain*. Wiley: Chichester, 2004.
[5] Jason Bloomberg *et al. Service Oriented or Be Doomed!* Wiley: Chichester, 2006.
[6] Jonathan P. Castro. *All IP in 3G CDMA Networks*. Wiley: Chichester, 2004.
[7] Jeffrey Bannister *et al. Convergence Technologies for 3G Networks – IP, UMTS, EGPRS and ATM*. Wiley: Chichester, 2004.
[8] www.TMForum.org

5

NGN Operator, Provider, Customer and CTE

Until now, telecom network operators have been further divided into the fixed network operator, the mobile network operator and the cable network operator. These operators provide network-associated telecom services and broadcasting services and therefore are also service providers.

However, there also exist pure service providers who do not operate a physical network, typically ISPs for IP services and MVNOs (mobile virtual network operators) for mobile telecom services.

The customers are individuals and corporations, consuming fixed and mobile telecom services, IP services and broadcasting services via different CTEs.

- The CTEs for individuals are fixed phones, mobile phones, PDAs (personal digital assistants), laptops, desktop PCs, televisions, radios, etc.
- The CTEs for corporations are PBX and corporate servers, where PBX enables corporate telecom services like short number dialling and corporate servers enable corporate IP services like intranets.

In an NGN environment, the open interface-based service and network architecture will enable an enormously extended landscape for network operators, service providers and customers. In principle,

- An NGN network operator is not necessarily an NGN service provider.
- An NGN network operator is not necessarily an end-to-end network operator; it can be an access network-only operator, or a transport network-only operator.

Next Generation Networks: Perspectives and Potentials Jingming Li Salina and Pascal Salina
© 2007 John Wiley & Sons, Ltd

- NGN services are networked IT applications, including the telecom, broadcasting and IP services of today where traditional telecom services are simulated or emulated.
- An NGN service provider can also be an end-user or a third-party provider.
- An NGN customer can be an individual, a corporation, a home or a vehicle. A third-party service provider can also be considered as a special customer that consumes the basic network services like call set-up and location information to provide advanced applications to end users.
- NGN CTEs will be all based on ICT where the IT is for services and the CT for connectivity.

5.1   NGN NETWORK OPERATOR

The NGN operators enable the *transport stratum* in the NGN architecture given in Figure 4.2. The two open interfaces within the *transport stratum* are the interface between the *transport control layer* and the *transport layer*, and that between the *access transport network* and the *core transport network*. These open interfaces enable the very fragmented NGN operators.

Of course, a full NGN operator that takes responsibility for providing the required end-to-end connectivity according to the *service stratum* provides the *transport stratum*.

However the *transport stratum* can also be provided by a group of operators which is composed of a single *transport control layer* operator and many *transport layer* operators, including:

- access network-only operators that can be access technology specific, e.g. satellite access operators, UMTS terrestrial access operators, ADSL access operators, WiMAX access operators and WLAN access operators.
- core transport-only operators that can be core transport technology specific, e.g. ATM transport operators, MPLS transport operators and GMPLS transport operators.

When an end-to-end connectivity is set across the networks provided by different operators, the performance is controlled by the *transport control layer* and guaranteed by the SLA bounded between the *transport control layer* operator and the *transport layer* operators. The types of fragmented NGN operator are:

- Transport control-only operator – such an operator provides centralized NGN core control functions, including

  o managing the access and the core transport networks from different operators, e.g. new operator registration, old operator removal, performance monitoring and load monitoring;

    ○ setting up, maintaining and tearing down end-to-end connectivity according to the requirements from the service stratum and according to the user terminal capability, the availability and load of access and core transport networks from different operators;

    ○ supporting different types of end-to-end connectivity for

        (a) point-to-point or point-to-multipoint or multipoint-to-multipoint connectivity;

        (b) simplex or duplex traffic;

        (c) session-oriented or non-session-oriented traffic;

        (d) stationary terminals, nomadic terminals and on-the-move terminals;

        (e) simultaneous usage of multiple terminal for a single service, e.g. television for downlink traffic or mobile phone for uplink traffic;

        (f) single terminal for bounded services, e.g. *call conference* and *we share* at the same time.

• Access transport-only operator – such an operator can be

    ○ a satellite access network operator;

    ○ a VDSL access network operator;

    ○ a mobile access network operator;

    ○ a WiMAX access network operator;

    ○ a WLAN access network operator.

and such an operator is capable of

    ○ providing access connectivity with the performance required by the transport control;

    ○ sharing access capacity among multiple transport control operators.

• Core transport-only operator – such operator can be

    ○ a MPLS based transport operator;

    ○ a GMPLS based transport operator;

    ○ an ethernet based transport operator;

    ○ an ATM based transport operator.

such an operator is capable of

    ○ providing the core connectivity with the performance required by the transport control;

    ○ sharing the core transport capacity among multiple transport control operators;

    ○ interconnecting with multiple access networks based on different access technologies;

    ○ interconnecting the core transport network with other core transport networks based on different transport technologies.

5.2　NGN SERVICE PROVIDER

In an NGN environment, enabled by the open interface between the *service stratum* and the service/application/content/information layer, there will be numerous providers for service, application, content and information. Furthermore, enabled by the decoupling of the *service stratum* and the *transport stratum* and the flexible IP-based networking capability, a service consumer can also be a service provider for the targeted audience. That is, in addition to the traditional service provider, an NGN service provider can be any individual, any home, any organization or any enterprise.

- *Service providers* – such providers provide NGN IP-based services like email, remote access, WWW browsing or VoIP, and also the simulated or emulated telecom services of today, like voice calls.
- *Application providers* – such providers provide applications like e-government, e-commerce, e-learning, e-voting and e-medicine.
- *Content providers* – such providers provide television, video, music and games like Second Life.
- *Information providers* – such providers provide news, traffic information, natural disaster forecasts and other information.

In order to deliver the service/application/content/information from anyone to anyone in a guaranteed way, a well-designed service delivery platform embedded in the *service stratum* is essential.

5.3　NGN CUSTOMER AND CTE

NGN is designed to serve a broader customer base, including individual customers, corporate customers, home customers, vehicular customers and third-party providers. Different customers are served with different CTEs. In general, the NGN CTEs will all be ICT-based and with integrated capability for IT service and CT connectivity.

5.3.1　Individual Customers and CTEs

The NGN individual customer holds the identities of personal phone number, personal IP address, personal SIP address, personal email address and personal ULR. The CTEs for individual customers are mostly mobile ICT devices, e.g. mobile phones, PDAs, multimedia computers and laptops, and of cause also fixed terminal equipment, e.g. television sets and desktop PCs.

Those mobile individual CTEs can become part of a home network to which they are registered, once they have moved into the coverage of the home network. Depending on the CTE's capability, the individual customer consumes end-user-relevant service/application/content/information possibly with different quality levels, e.g. high-definition (HD) video or standard-definition (SD) video.

5.3.2 Home Customers and CTEs

The NGN home customer holds the identities of home phone number, home IP address, home SIP address, home email address and home ULR. The home customer pays a home network, which is a local network for the home environment. Home electronic devices and a home gateway form the home network. Home electronic devices can be:

- Non-portable electronic equipment including television, desktop PC, printer, refrigerator or monitoring camera. These devices are identified with the home they belong to, and are capable of local connectivity, wired or wireless. The wired connectivity can be provided by ethernet, PLC (power line communication), etc.; the wireless connectivity can be provided by InRD, WLAN, Bluetooth, UWB (Ultra WideBand), etc. When new equipment arrives, it can listen to the home gateway, register to it and become part of the home network. As long as those registered devices are switched on, they are detected by the home gateway automatically and become part of the HomeNetwork. Before old equipment is eliminated, it should de-register from the home gateway.
- Portable electronic devices including cameras and PM3. These devices are identified with their owner and capable of local wireless connectivity through NFC, InRD, Bluetooth, UWB, WLAN, etc. The devices can register to the LAN they can hear to be connected to the Internet. When switched on at home, they can be detected by the home gateway and become part of the home network.
- Mobile devices, including mobile phones, laptops and multimedia computers. These devices are identified with their owner, and capable of mobile system connectivity and local wireless connectivity. As soon as such devices are moved back home and switched on, they will be automatically detected by the home gateway and included in the home network, i.e. they are ready to use the local connectivity to carry out telecommunications with the advantage of higher reliability, higher quality and lower price.
- Vehicles, including cars and vans. These vehicles are identified with their owner and are equipped with tailored mobile communication capability, close distance communication capability and local connectivity capability, where

  o the tailored mobile communication capability includes, for example, SMS for accident or unusual machine status warning and real-time information downloading for driving route optimization;

o the close distance communication capability includes, for example, UWB-enabled car-to-car communication for automatic navigation;
o the local connectivity capability via WLAN, Bluetooth, etc. is automatically included in the home network to enable automatic machine control according to the instructions from the garage; reports are received from and sent to the garage via the home network when at home.

The home gateway administers the home network and is an ICT device itself. It holds the identities of home phone number, home IP address, home SIP address, home email address and home ULR.

The home gateway is connected to the external network via single or combined access technologies like twisted pair, cable, fibre, fixed wireless, mobile or satellite. It is connected to the home network elements via local connectivity.

• As an ICT device, the home gateway itself is a home phone with PBX functions, which

   o is reachable by home phone number, home IP address or home SIP address;
   o is capable of all kinds of telecom services e.g. voice call, video call, SMS or MMS;
   o is connected to the message box when line is busy;
   o can program a home-specific welcome message;
   o can program a short number for the personal phone of each family member; by dialling a single number, the call is re-dialled to the personal phone, e.g. a mobile phone when at home.

• As an ICT device, the home gateway is a home portal, which

   o is reachable by home IP address, home SIP address or home URL;
   o is capable of all kinds of IP-based communication services, e.g. VoIP click-to-call, MMoIP click-to-call, instant message and email;
   o acts as the set-top box for IP-based broadcasting, multicasting and unicasting services;
   o can design and display a home-specific WWW page with welcome message, introduction to the family member and actual information about the family and family members. By clicking the picture or name, a call or a message to this person can be initiated; family videos or other information can also be viewed by authorized users.

• As an administrator, the home gateway:

   o is a relay which enables the telecommunications capability of home electronic devices equipped only with local connection capability;

- o is a short-cut enabling higher reliability and quality as well as lower cost connectivity for telecommunication-capable devices;
- o manages a single communication service via multiple home devices, e.g. television for downlink video and phone for uplink video or voice;
- o manages the direct connection between devices within the home network;
- o manages the inventory of home equipment and devices;
- o sends signals for automatic home management, e.g. when the washing is finished;
- o sends warnings to the home owner when anything unusual is detected at home;
- o enables the broadcasting service for home HD television with interactive capability controlled via mobile device.

5.3.3 Vehicle Customers and CTE

The NGN vehicle customer holds the identities of vehicle phone number, vehicle IP address, vehicle SIP address, vehicle email address and vehicle ULR. The vehicle customer pays for a vehicle LAN, which is a moving LAN. A vehicle LAN can be a CAN (car area network), BAN (bus area network), TAN (train area network), SAN (ship area network), PAN (plane area network) or any ad-hoc LAN within a moving body.

A vehicle LAN is composed of a vehicle gateway, sensors and electronic devices inside the vehicle. The vehicle gateway administers the vehicular LAN and is an ICT device itself. The vehicle gateway is connected via single or multiple wireless access technologies like mobile or satellite.

- As an ICT device, the vehicle gateway which

  - o is reachable by vehicle phone number;
  - o is capable of all kinds of telecom services, e.g. voice call, video call, SMS or MMS;
  - o is connected to the massage box when the line is busy.

- As an administrator, the vehicle gateway:

  - o stores all the information about the vehicle (type, usage, accident record);
  - o collects current status about location and driving speed;
  - o sends the relevant information about the vehicular regularly or on request;
  - o collects and processes the information provided by the sensors to enable automatic driving, automatic parking and driving safety (when the sensor detects driving too fast or too close to another vehicular, it warns the driver and communicates to the vehicle in danger or causing danger);

- o manages intelligent driving, i.e. navigates according to received road information and availability to chose the shortest way in time or in distance;
- o initiates a warning when anything unusual is detected or sensed at the vehicle;
- o manages the direct connection between devices within the vehicle network;
- o manages the inventory of equipments and devices within the vehicle.

5.3.4 Corporate Customers and CTE

The NGN corporate customer will be served with an IP-PBX solution, which merges the separated corporate voice and data services of today, such as voice via TDM-PBX, data or IT services via intranet. The IP-PBX solution serves both the employee internally and the customer externally with integrated TDM-PBX-like functions and portal-like functions, where

- the TDM-PBX-like functions for the employee incorporate:
  - o private numbering plan with short number dialling;
  - o address book with presence and availability information;
  - o any terminal including mobile phone, fixed phone, IP phone, laptop, PC;
  - o optional special tariffs for all calls;
  - o special tariffs for internal calls;
  - o black and white list for outgoing calls;
  - o supplementary services such as call transfer, multiparty or conference call.
- The TDM-PBX like functions for the external customer incorporate:
  - o hunting group – the enterprise/organization can define a hunting list, a series telephone numbers. If the first number is not reachable, the call will be routed automatically to the next number according to the defined list;
  - o automatic switchboard – when calling the enterprise number, the caller is asked to make a choice according to the predefined single- or multi-level selection, e.g. push 1 for sales department, push 2 for service department;
  - o queuing incoming calls – if the switchboard attendant is busy, the caller hears an announcement and is put on hold in the queue until his or her turn;
  - o immediate call transfer – when the caller wishes to transfer the call to another party, the call is put through immediately without the OK from the previous called party.

- The portal-like functions for the employee incorporate (intranet):

  o web pages for internal information;
  o web interface for internal communication among employees with location, presence and availability information, e.g.

    (a) click-to-call (voice, video, text) – by clicking at the name or picture to initiate, for example, an on-line discussion, an ad-hoc meeting or a conference call; the attendants could be physically located at anywhere in world;
    (b) on-line document edit – with shared board, to discuss and edit document or presentation slides on-line together.

  o corporate IT environment access (document download/upload, scheduler), e.g.

    (a) an employee at any place (office, home, on-the-move) could organize meeting according to the availability of persons and room;
    (b) an employee at any place (office, home, on-the-move) could fetch needed corporate information for remote demonstration or remote ad-hoc meeting.

- The portal-like functions for the external customer incorporate (Internet):

  - a web site for the information on the enterprise or organization;
  - a webinar for promotional information of the enterprise or organization;
  - clicking the product or service to talk to the person concerned;
  - monitoring the web-page utilization rate, monitoring the company documents traffic via web page or via FTP server.

- The administrator for the corporate resource management provides

  o advanced employee management with location information, for example, coordination of work with teleworking or on-the-move employees.
  o advanced vehicular management with location information, for example, vehicular status or location.

5.3.5 Third-party Provider Customers and CTE

The third-party provider is a special type of customer, which utilizes the NGN network service enablers to create and provide advanced service/application/content/information. Depending on the operator, the available service enablers can be different, including for example,

- call control (call set-up, manipulation of multimedia);
- data session control (data service);

- user interaction (voice and data exchange between application and end user);
- generic messaging (SMS, MMS, instant messaging);
- connectivity management (QoS);
- mobility (user location and status info);
- terminal capability;
- user account access and management;
- policy management;
- presence and availability management;
- location information;
- conference call.

The CTE of a third-party provider includes an IT server and database. The advanced application can be:

- give me a call when friends are within 50 m;
- alert me when a special store is near by.

6

Network and Service Evolution towards NGN

Today, several types of physical communication networks are running in parallel. Each type is associated with its native services and is designed optimally for its associated native services, e.g.

- The fixed telecom network is primarily designed for the voice communication service with twisted pair to the fixed phone.
- The mobile telecom network was originally designed for voice communication service enhanced by mobility.
- The cable network was designed to provide broadcasting services television and radio with coax cable to the television set or radio receiver.
- Internet was designed to provide the basic Internet services of web browsing, email, FTP and remote login above a fixed telecom network.

However, with the advances in technology and the competitive environment for customers, this landscape of networks and services has been broken up and is changing rapidly.

Even though the changes are different from network to network and different from service provider to service provider, the trend of changes is identifiable and can be summarized as follows:

- Besides the native services, each type of legacy network is trying its best to provide the basic Internet services like Web browsing, email, FTP and remote login with the move to provide advanced Internet applications like VoIP and IPTV.

Next Generation Networks: Perspectives and Potentials Jingming Li Salina and Pascal Salina
© 2007 John Wiley & Sons, Ltd

- Each type of service provider is trying its best to enhance its capabilities and features to the maximum extent to provide the services of others, like the telecom network providing broadcasting services, the ISP providing telephony services and the cable network providing Internet services.
- The network feature and capability enhancements include bandwidth increase, wireless access and mobility support.

All these moves have signalled the beginning of the network evolution towards NGN.

The NGN architecture will harmonize the fragmented situation of network and service, to prosperous services, applications, contents and information for end users, to make the best use of the existing network infrastructure and as well opening new access and core network technologies. It brings the advances of:

- Network-agnostic access to services, applications, content and information. The serving network will be chosen based on the customer subscription, service requirement, terminal capability and network resource availability.
- Utilizing all existing access networks as the service access channels for pervasive service accessibility.
- Being capable of integrating new access networks for the enhancement of access capability.
- Utilizing all existing transport networks as possible service carriers.
- Being capable of integrating new transport networks for the enhancement of transport capability.
- Making the best use of co-existing access and core transport networks in order to utilize the network resources according to the requirements from the *Service stratum.*
- Enabling a manageable end-to-end connectivity satisfying the QoS and security requirements.
- Delivering and managing services via a common platform that is agnostic to the underlying transport network.

However, there is still a long way to go to reach this harmonization of NGN from today's messed-up situation of network and service. In principle, all types of active players today have the chance to finally become an NGN operator or a provider. The successful ones will be those who follows the right evolution strategy.

To design a wise evolution path, it is necessary to understand each key evolution step in terms of *what this step is doing* and *what does this step mean for the final goal.* Moreover, if the final goal can be reached without going through some or even all of the intermediate step(s), one should skip them without hesitation.

6.1 MAJOR EVOLUTION STEPS FOR THE NETWORKS AND SERVICES OF TODAY

We see three major evolutionary steps towards NGN from today's networks and services, which are:

Step 1: service convergence and access network development.
Step 2: IP-based service conversion and managed IP network development.
Step 3: network integration and service extension

It is not necessary to follow the steps in strict sequence – the next step can start before the previous one has finished.

The evolutionary Steps 1 and 2 can be started and progress within a fixed network domain or a mobile network domain or a cable network domain. However, the accomplishment of Step 3 must be done across all three physically existing networks. Today, we have already taken the first step.

6.1.1 Service Convergence and Access Network Development (Step 1)

The *service convergence* serves the customer with the same service from different networks, as shown in Figure 6.1, where a customer can watch the same television programme at a PC via PSTN network, at a mobile phone via a mobile network or at a television set via cable network television.

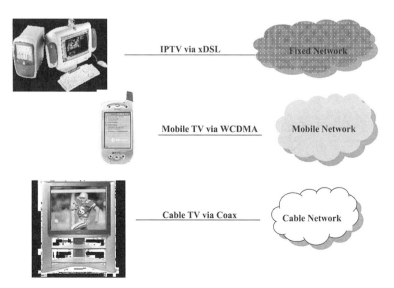

Figure 6.1   An example of service convergence: television is available via xDSL at a PC, via WCDMA at a mobile phone and via Coax at a television set

For an existing network, such *service convergence* means being able to provide the foreign services associated with other types of networks. The technology for the service realization could be different. The delivered QoS could also be different.

Considering that each service has its QoS requirement on the network, including bandwidth, delay, jitter, data error rate, etc., to cooperate with such *service convergence*, the capabilities of an existing network need to be enhanced, specifically the access network which is usually the bottleneck.

The *access network development* includes the aspects of:

• increased access bandwidth;
• reduced delay;
• support for the portability and mobility (from low- to high-speed moving terminals).

6.1.2   IP-based Service Conversion and Managed IP Network Development (Step 2)

The *IP-based service conversion* will substitute the native services associated with legacy networks with IP-based ones, e.g. with VoIP for telephony. This conversion decouples the services from the underlying transport networks, which is the key to realizing access-agnostic service accessibility. This is shown in Figure 6.2, in which a laptop can access the same television programme via fixed, mobile and cable access via an adequate adapter; furthermore, this decoupling enables services to be added or removed without touching the underlying networks, and the changes in the underlying network have no impact on services.

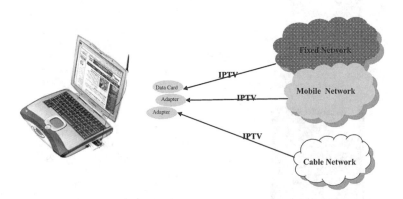

Figure 6.2 An example of IP-based service conversion: IPTV available at a laptop via xDSL, WCDMA and Coax

It should not be a simple one-to-one service conversion. Enhanced features should be added to the converted services, e.g.:

- The traditional telecom voice service is converted into VoIP or MMoIP (multimedia over IP), which will be available via fixed, mobile and cable networks, with the possible enhancement of click-to-call, group call, address book with presence and availability information and adding/removing media according to need and terminal capability.
- The traditional cable television broadcasting service is converted by IPTV, which will be available via fixed, mobile and cable networks, with the possible enhancement of interactive television, unicasting/ multicasting/broadcasting and consumer provided content.

As we know, a legacy network is designed and optimized for its native service. The QoS and security required by the native service are built into the network through protocols, algorithms and switching speed. After being decoupled from its associated service, the legacy network is there to carry any IP-based services that could have very different QoS and security requirements, the previous built-in QoS and security measures can be no longer sufficient. That is to say, new mechanisms have to be found and implemented to transform such a network into one of many underlying networks, which are capable of providing different levels of QoS and security according to the service requirement, a so-called managed IP network.

The *managed IP network development* will transform the best-effort IP network of today with legacy networks into a manageable IP network for delivering IP-based services with required quality and security. This development includes:

- unifying the language of QoS requirement description independently from underlying networks;
- embedding QoS mechanisms along end-to-end IP connectivity chain, e.g. admission control, scheduling or congestion control;
- being able to activate the embedded QoS mechanisms along the chain of end-to-end IP connectivity in a consistant way, e.g. the core network will not give a data pipe higher than what the radio access network is ready to give;
- unifying the language of security requirement description independently from underlying networks;
- embedding security measures along end-to-end IP connectivity chains;
- being able to activate the embedded security mechanisms along the chain of end-to-end IP connectivity in a consistent way, to avoid unnecessary duplication of security, which usually lowers system performance.

6.1.3   Network Integration and Service Extension (Step 3)

The *network integration* will add a common *transport control layer* above all the access and core transport networks, forming the *transport stratum* of NGN architecture, as shown in Figure 6.3.

The *transport control layer* is responsible for utilizing all types of access and core transport networks when setting up and maintaining the end-to-end connectivity according to the service requirement.

After experiencing Step 2, a legacy network becomes a pure carrier for IP packet transportation end-to-end or part of end-to-end, with the capability to handle the traffic differently according to the requirements for QoS and security. Under this common *transport control layer*, the traditional fixed, mobile and cable networks are integrated into a common transport pool to be called to form an end-to-end connectivity with the required QoS and security. In addition, the formed end-to-end connectivity is maintained by the management functions also located in the *transport control layer*.

The *service extension* is to place the *service stratum* and the open interfaces to enable the creation and delivery of prosperous services for end users. Unifying similar functions from fixed, mobile and cable networks can form the service stratum. This unification should be understood in the logical sense, i.e. the same

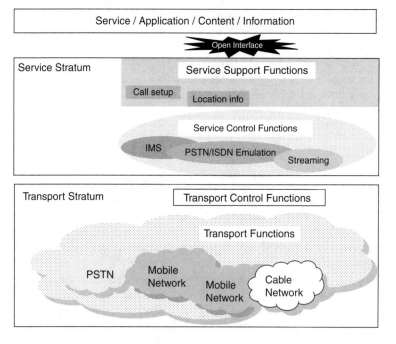

Figure 6.3 NGN *transport stratum*, *Service stratum* and open interfaces towards service

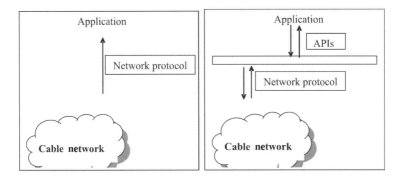

Figure 6.4 (Left) service creation without open interface; (right) service creation with open interface

types of physical entities for fixed, mobile and cable networks will remain but under a centralized control. A common *service stratum* is shown in Figure 6.3, where:

- the *service support functions* should include rich network service enablers like call set-up and location information to enable the creation of advanced services, applications, content and information;
- the *service control functions* should include the components of IMS, PSTN/ISDN emulation and streaming to guarantee service delivery to end-users with sufficient quality and adequate security.

The open interface, between the *service/application/content/information layer* and the *service support functions*, is for the easy and standardized creation of service, application, content and information. Without an open interface, it could take years to develop a service because a service developer has to master the network protocols, which are usually very complex and vendor-proprietary, as shown in Figure 6.4 (left). The open interface, defined by a group of APIs and used to call the network services in a high level language like Java, standardizes and simplifies the service creation greatly, as shown in Figure 6.4 (right).

There already exist several groups of APIs, e.g. OSA/Parlay APIs and Parlay X APIs. Take the OSA/Parlay APIs as the example, which include:

- *Call control* for setting up and maintaining a basic call;
- *User interaction* for exchanging data between an application and an end-user;
- *Mobility* for obtaining the location and status of an end-user or multiple end-users;
- *Terminal capabilities* for obtaining the capabilities of an end-user terminal;
- *Data session control* for setting up and maintaining a data session;
- *Account management* for accessing the accounts of an end-user;

- *Content-based charging* for charging on-line the end-user for the usage of service and/or data;
- *Generic messaging* for accessing the messaging box of SMS, multimedia messaging, etc.;
- *Connectivity management* for providing provisioned QoS;
- *Policy management*;
- *Presence and availability management.*

In this step, the final architecture of NGN is formed, and the advanced NGN management is in place. The NGN transformation is finished from today's point of view. The formed NGN architecture is an SOA (service-oriented architecture) and will enable the services to be extended far beyond the simple addition of today's telecom, broadcasting and IP services. The first will be:

- e-applications including e-commerce, e-goverment and e-learning;
- home networking including remote and automatic control and virtual home;
- networked sensor applications including traffic navigation and natural disaster forecasting.

6.2   FIXED NETWORK EVOLUTION

The fixed telecom network PSTN has evolved mostly from its original PSTN design for telephony services with twisted pair, to ISDN and to ADSL. Under the constant demand for broader bandwidth Internet access, mobility and home/office networking, the evolution continues in the *access network domain* with the moves to:

- introduce accessibility of hundreds of Mbps with optical fibre to home or to building, for real-time IP multimedia services like HD IPTV;
- introduce wireless accessibility of hundreds of Mbps for tens of metres coverage with ultra wide band or wireless LAN or WiMAX, for nomadic mobility at home or in the office;
- introduce connectivity of hundreds of Mbps between rooms within a home or an office with Power Line Communication, for home or office networking.

With the demand for the managed IP transport network, the evolution continues in the *IP transport network domain* with the moves to:

- carrier-grade ethernet network;
- IP transport network capable of fine QoS granularity with DiffServ, MPLS, GMPLS and MPLS-T;
- IP transport network capable of different security levels with IPsec, etc.

The demand for *service substitution* with IP-based ones results in:

- extension of the IP from backbone network towards end-user devices to enable the communication services to be substituted with VoIP, MMoIP or IM (instant messaging);
- extension of the IP from backbone network towards enterprises and organizations to enable IP-PBX-based services for the corporate environment or home or office.

The demand for *service extension* leads to:

- the introduction of IMS for SIP-based peer-to-peer communication services;
- the taking over of the mobile access and cable access to form a common IMS in the logic sense by interconnecting the same type of physical entities of IMS from the fixed, mobile and cable IMS domains, e.g. forming a single logical HSS by interconnecting fixed HSSs, mobile HSSs and cable HSSs [1].

6.3 MOBILE NETWORK EVOLUTION

The mobile telecom network, dominated by GSM, has evolved most rapidly from its original design for mobile voice service with lower speech quality, to GPRS and later EDGE, which enable the mobile data services of narrow band, to UMTS and recently HSDPA and HSUPA under the requirement for a broadband mobile data service like mobile television.

It is a well accepted fact that the capability of mobile networks is about 5 years behind that of fixed networks. Based on this fact, mobile network evolution is rather easier to define. In the *radio access domain*, the next moves will be:

- introducing Mbps uplink capability with HSUPA, starting from 2007;
- introducing tens of Mbps for both uplink and downlink with LTE (long-term evolution), starting from 2010;
- introducing hundreds of Mbps for downlink with the next-generation mobile system or 4G or IMT-advanced, starting from 2015.

In the *core network domain*, the next moves will be:

- introducing flat IP transport core for 10 ms user data latency from end to end with SAE (system evolution architecture), starting from 2010.
- introducing interworking capability with roaming and handover support for service continuity among major co-existing mobile and wireless access systems like WiMAX and WLAN.

In *backhaul network domain,* the next moves will be:

- introducing meshed multiple radio hops;
- introducing EPON.

In the *service domain*, the next moves will be:

- introducing IMS for mobile SIP-enabled peer-to-peer communication services;
- taking over the fixed access and cable access to form a common IMS in the logic sense.

6.4 CABLE NETWORK EVOLUTION

The cable network is the last to evolve from its original design of analogue television and radio broadcasting services. The on-going moves are towards digital television and radio broadcasting services, and Internet services. The next move will be towards interactive television, HD television, large-screen digital imagery television and time-shifted narrowcast television. On the other hand, there are also moves to provide IP-based services like Internet services and VoIP service.

6.5 INTERNET EVOLUTION

The Internet, operated by an ISP, does not have its own physical network, instead sitting on the fixed telecom networks to provide basic Internet services of email, web-browsing, FTP and remote login, and also advanced IP applications like VoIP and IPTV.

Such an ISP does not have a direct customer base. However, this situation could change rapidly with various wireless access technologies that allow the rapid deployment of an end-to-end IP network or an IP access network at least. Under such circumstances, the first wireless ISPs will come out very soon. For example, a WLAN-based wireless mesh network (IEEE802.11s) enabled ISP can provide its customers with broadband access and nomadic mobility.

The next moves will be:

- to use Web 2.0;
- to use IPv6;
- to provide advanced IP-based communication services with enhanced features and at least equivalent QoS and security to substitute for the communication services provided by telecom networks until now;

- to provide advanced IP-based broadcasting services with enhanced features and at least equivalent QoS and security to substitute for the broadcasting services provided by cable networks until now.

6.6 IP NETWORK PROBLEMS CRITICAL TO BE SOLVED

The picture about NGN is beautiful. However, based on all IP for both services and network, there are still critical issues, especially located on the network side from the all-IP concept. Network operators need to work out those issues first before rushing to the next step; otherwise the evolution will only lead to the frustration of end-users.

In order to carry the NGN all-IP based services in the way they are supposed to be carried, underlying networks need to be optimized to transport the IP packets for mixed services with various traffic characteristics and requirements for QoS and security (for both signalling and media data).

Considering today's underlying network capability for the end-to-end IP packet transportation, the problems can come from:

- *Private IP addresses* hurdling the real-time IP services. Owing to the lack of IP addresses from the IPv4 address base, private IP addresses are used. The end-to-end IP connection between those private IP addresses is realized via a NAT (network address translation) or many NATs. Such end-to-end IP connection via NAT(s) hurdles the IP peer-to-peer real-time communication. Therefore IPv6 should be deployed as soon as possible.
- *Inefficient usage of radio resources* for IP over mobile and wireless access. The voice-like real-time traffic generates short IP packets. To transport such short IP packets, the IP header (+ UDP + RTP) can triple the size of the packet. According to a rough estimate, there will be only ~30 % efficiency for VoIPv4, and ~20 % for VoIPv6. These extra bytes also contribute to increased delay. Considering the scarcity of radio resources, such heavy overhead is inefficient and not acceptable. The possible solution is header compression at least over the radio access part. 3GPP is working on the topic for the mobile access; however the IEEE does not consider such topic for wireless access technologies like WLAN or WiMAX. Therefore IP header compression should be applied, for example cRTP can compress the 40 byte IP/UDP/RTP (20/8/12 byte) header of a VoIP packet down to 2–5 bytes per packet.
- *Much longer call set-up time* for VoIP. The IP services are implemented with multilayer protocols, softswitch and intelligence; the call set-up time is

much longer than the PSTN network (estimated ~ 0.75 s for a PSTN call; ~ 20 s for VoIP).

- *Non-common QoS language* that could guarantee the understanding of the QoS requirement between interconnected IP networks. For some time, for a private IP network, the QoS has been guaranteed to a certain extent via over-dimensioning. However, this is not the case for a public IP network, which crosses many IP networks. The reason for this is that there is no a common QoS language for IP networks, i.e. the QoS markers from an IP network cannot be understood by another IP network; therefore measurements must be taken to ensure packet transport according to the marker. This is essential to realizing the manageable IP network required by NGN.

- *Inappropriate re-transmission concept* of Transmission Control Protocol (TCP) for wireless connection. The TCP tracks the datagram and dictates the retransmit when packet loss happens. However, TCP protocol is designed for the good performing network, e.g. optical networks. When the packet is not received correctly in a defined time frame, the TCP assumes that network congestion has happened and thus reduces the speed of sending data aiming to release the network from congestion. When using TCP for the network containing wireless connections, e.g. a mobile network, there are problems in assuming network congestion and cutting the data sending speed when a packet is not received correctly within a defined time frame. This is because, in a mobile network, when a packet is not received correctly within a defined time frame, in most cases, it is not due to network congestion, but rather to the poor radio connection or additional delay caused by the re-transmission at the data link layer or handover or cell reselection. To slow down the sending speed by TCP can only worsen the case. This has happened to a mobile operator: the physical channels were proven to be present, but the throughput of TCP traffic was zero. It was caused by a TCP confusion that had cut the data sending at the source port, therefore nothing was received at the receiver side. Therefore, to increase the TCP performance on radio link, the conventional TCP configuration needs to be adjusted for TCP window size and segment size by considering the round trip time and its variance.

- *Non-guaranteed end-to-end QoS* for real-time or multimedia IP service due to non-standardized mapping rules and effective control mechanisms across different IP networks. MPLS will improve the situation; however MPLS will provide relative QoS, far from guaranteed QoS as required. Also the mapping of QoS parameters to the label is not standardized. IPv6 is expected to solve the problem. However, TCP assures only that the data is delivered correctly; no guarantees are given against delay and throughput.

- *Slow introduction of IPv6* [2]. It has been clear from day one that the introduction of IPv6 will be stepwise and there will be no a Flag-day as there was with the introduction of IPv4. However, the pace is still much slower than expected. The NGN fundamentals of ubiquitous networking and pervasive accessing demand a tremendous address base, where IPv6 is mandatory to guarantee an IP address for every grain of sand on the earth. Besides the huge address base, IPv6 also means

○ peer-to-peer real-time communication, including person-to-person, person-to-machine and machine-to-machine without NAT in between to retard the direct communication between two IP addresses;
○ auto-configuration for network management;
○ efficient routing and processing using fixed header length and simplified header format;
○ improved support for extensions and options with extension headers;
○ end-to-end security support with IPsec for encryption or cryptographic authentication, and privacy. IPv6 has the key value of many new automation features that reduce operational overheads. However, inherent with increased automation are increased vulnerabilities, e.g. malicious users can spoof solicitation, advertisement and binding messages. To combat this problem, mechanisms have been developed to provide securer automated capabilities. Given the current instability of IPv6 relative to IPv4, it is expected that many security 'holes' are likely to be found in IPv6, especially as it continues to be deployed. However, as IPv6 evolves, the robust security features associated with IPv6 will improve because of increased use of end-to-end security functions. In the near-to short-term, it is anticipated that IPv6 will deliver minimal value addition to security over what is realized in the IPv4-only network today.
○ Simplified mobility support with mobile IPv6, which supports better the mobile device applications than IPv4.
○ Improved QoS capabilities with introduced flow label in the header.
○ IPv4 treats all the bearer traffic as being aggregated and does not distinguish between flows. Flows in IPv4 are typically identified using a tuple of packet header information including source and destination addresses and port numbers. Many application protocols use dynamic port allocation, which makes flow identification difficult in IPv4.
○ IPv6 headers provide for improved flow identification (with the definition of a flow ID field), but no standard mechanisms are yet in place for the use of this information. A single value in the flow ID field could enable the network to identify as a single coherent flow what would appear to be multiple flows in an IPv4 network; e.g. the control and data channels used for an FTP transfer.
○ Deployment of a new class of applications that is critical to e-business growth.

With these merits, IPv6 is enabling the Internet to go beyond the information sharing means, to become the platform for virtual organization, grid computing, home networking, networked sensors, amongst other things.

REFERENCES

[1] 3GPP TS 22.228, Service Requirements for the Internet Protocol (IP) Multimedia Core Network Subsystem.
[2] http://www.ipv6forum.com

7

NGN Key Development Areas

Today, we are in a time of technology explosion. However, in realizing the fundamentals of NGN, we are only at the start. There is much to develop, from technologies to methodologies, before the realization of NGN. The key developments can be summarized into the following areas:

- terminals;
- access networks;
- backhaul networks;
- core transport;
- service creation;
- network management;
- service management.

In Figure 7.1, each area is marked in the NGN architecture.

In the following section, we explain the fields to be developed and the promising technologies to meet this need.

7.1 TERMINAL AREA

Today, terminals mean to us user terminals, e.g. mobile and fixed phones. In an NGN environment, the terminal is extended from *user terminal* to *machine terminal* and *sensor terminal*, enabling connection between user and network, between machine and network, and between sensor and network. A new type of wireless terminal, the *wireless thin client*, is also emerging. Each type of terminal has its own development dimension and space.

Next Generation Networks: Perspectives and Potentials   Jingming Li Salina and Pascal Salina
© 2007 John Wiley & Sons, Ltd

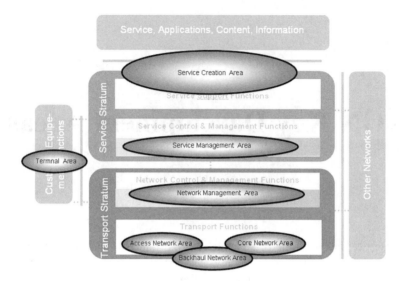

Figure 7.1 The key development areas marked upon the suggested NGN functional architecture

7.1.1   User Terminal

A *user terminal* is the interface for users to access the service/application/ content/information via the network, and can also be considered as the first network node. However, the terminal interface of today is primitive, not designed for the convenience of humans, who are actually reluctant to use the service.

The NGN promises that the *user terminal* will be extremely simple to use and be always connected to the network, which demands that the NGN *user terminal* to be *highly intelligent, highly flexible* and *highly adaptive.*

Highly intelligent means that a user should interact with the terminal in a more human-like way, especially for human-to-human communication, e.g.

- authenticating the user with personal biometrics, e.g. face, veins, figure prints or eyes, instead of typing codes or passwords [10];
- starting a conversation by calling the person's name instead of dialling a number or clicking at the name;
- directly writing a text message on the screen instead of typing at the keyboard;
- sketching a funny drawing instead of choosing from a ready-made picture library;
- discovering or searching a service by sketching or speaking the key words instead of browsing

- enabled editing of text (marking text, correcting words), pictures (modifying, adding, deleting, changing colour), music (changing tone, adding instruments) and video (changing order, changing speed).

Highly flexible means for a user to choose to activate the media and context information according to the communication need, where:

- the rich communication media include text, voice, picture, video, smell, taste and touch;
- the context personal information includes mood, feeling and availability;
- context environmental information includes local language, local time, local weather (temperature, humidity, wind), local surroundings and local conditions (vibration);

Highly adaptive means to hide technology from users by being remotely reconfigurable for the requested service, for the available access technology and for the management, e.g.

- An SDR (software-defined radio) capable terminal can be re-configured to use different access technologies and different bands, including short-range radio technologies, wireless and mobile middle- and long-range radio technologies and broadcasting technologies, and can detect and report the existing access technologies to assist the network to set up end-to-end connectivity in an optimized way.
- The software of a terminal is upgradeable or a new client can be uploaded to enable the requested service/application/content/information.
- The modular construction of a terminal, e.g. an external screen can be added when watching television programme, an electronic paper can be added when reading or writing.
- A stolen terminal can be disabled remotely via a software upload from the network.
- The capability of a terminal to report QoS received at the terminal (downlink), which enables the network to adjust the downlink physical channel when necessary.
- The capability of a terminal to be integrated into a local network or to be used as one of the terminals with others for a single service, e.g. television as the receiving terminal, mobile phone as the providing terminal.

7.1.2   Machine Terminal

Compared with the NGN *user terminal* device, an NGN *machine terminal* is much simpler as no human-like behaviour interface is needed.

The NGN *machine terminal* can be:

- an integrated part of a machine (vehicle, camera, printer, refrigerator, wash machine) with tailored telecommunications capability, e.g. sending SMS, receiving software updates, etc. where a trigger has to be designed for automatically activating the embedded telecom service;
- a gateway to administer the traffic between the public network and a local network, either stationary or on-the-move, e.g. HAN (home area network) or VAN (vehicle area network, including planes, ships, trains, buses and cars).
- an administrator, e.g. virtual PBX, to manage the internal and external communications in a corporate environment distributed worldwide.

Today, there exists only individual development in this field. The vision of NGN in which anything can be connected to anything still demands prosperous technology development, which can be:

- to equip a portable electronic device with xSIM and tailored telecommunications capability, including message sending/receiving or file downloading/uploading, e.g. camera or music player, where the xSIM allows the portable device to access the network service and to be allocated and the tailored telecommunications capability allows the downloading of music, the uploading of pictures or videos or the upgrading of device software.
- to equip the terrestrial VAN gateway with xSIM, vehicle-to-vehicle communication capability, broadband connectivity, tailored communication service capability, sending/receiving messages, uploading/downloading files, etc., where

  - the xSIM capability allows a vehicle to be traced for location, driving speed and distance from other vehicles, enabling centralized traffic management for automatic navigation, safe driving and provision of road traffic information;
  - the vehicle-to-vehicle communication capability enables safe driving, for instance when two or more vehicles are too close to each other, the vehicles can talk to each other to avoid accidents;
  - the broadband connectivity allows devices under the reach of the VAN to connect to the external network via the gateway;
  - the messaging send/receiving capability allows remote control for the vehicle status, e.g. information about the vehicle can be sent when requested or the embedded sensor will initiate a message when unusual parameters are detected;
  - the file download/upload capability can be used for on-line maintenance, e.g. vehicle software can be upgraded and actual driving maps can be uploaded.

- To equip the space VAN gateway with xSIM, space vehicle-to-vehicle communication capability and real-time broadband telecommunications capability, e.g. real-time video transmission, where

  - the xSIM capability allows the space vehicle to be traced for height, location, flying speed and distance from other flying vehicles, to enable centralized traffic management in space for automatic navigation and safe flying;
  - the space vehicle-to-vehicle communication capability enables safe flying – when two or more vehicles are too close to each other, the vehicles can talk to each other to avoid an accident.
  - the real-time broadband communication capability will enable real-time monitoring video to be sent to the ground station when there is any doubt about the status of the vehicle;
  - when any unusual case is detected, early help can be organized from the ground, including controlling a plane in the hands of terrorists to avoid tragedies similar to 9/11.

- To equip the HAN gateway with IP PBX capability, the multiple broadband access capability towards the public network, e.g. wired VDSL access, wireless WiMAX access, mobile LTE access and multiple access capability towards home equipment and devices like WLAN, Bluetooth, ZigBee, UWB and PLC, where

  - the mobile and portable devices can join the HAN via the gateway when moving into the coverage of a home network for a better connection quality with a better price;
  - the PBX-like function will enable a modern telecom environment at home, e.g. when a family is called, the caller will be informed about the availability of the family members and can start a conversation by calling the family members name;
  - the IP server-like function will enable a web-based family information site;
  - the home device can be connected to the public network via its access technologies to the HAN gateway, e.g. autonomously sending an SMS when the embedded sensor in the washing machine detects an unusual parameter;

- To equip advanced PBX for the corporate environment (enterprise, organization or government) with an integrated capability for PBX-like services and Web-portal-enabled services, where

  - the PBX-like services are for voice-oriented communication, including call conferencing with media adapted to the terminal's capability and access availability;
  - the Web-portal service is for the IT services of intranet and Internet.

- To equip very-short-range wireless access, e.g. NFC, with mobile terminals to connect the electronic world with the telecom world, where

  - the NFC-enabled mobile phone will be able to communicate with any electronic device, platform and system – the electronic world can generate content and the telecommunications world can upload or download the content, e.g. video or television programmes;
  - the NFC-enabled mobile phone can also be used for remote authentication, enabling e-commerce, e-banking and e-governance and e-voting.

7.1.3 Sensor Terminal

A *sensor terminal* is a terminal capable of reporting sensory information from embedded sensors or picking up sensory information via a network. The NGN promises to provide context-aware communication where the context information is generated via various sensors, e.g.

- personal sensors for gathering information about

  - a person's location (which is inferred from that of their devices),
  - a person's activities (inferred from calendar and desktop information),
  - a person's biometrics,
  - a person's medical status;
- environment sensors gathering information about

  - room status (inferred from anonymous motion/sound detectors),
  - the presence of people and things,
  - congestion (inferred from pressure pads or camera images).

The NGN also promises to collect the local and global data to predict natural disasters, where the local and global data are collected via networked sensors. Such *sensor terminals* should be able to follow defined criteria, and send sensory information only when the data has an unusual status.

The technologies for *sensor terminals* are critical in realizing this NGN vision.

7.1.4 Wireless Thin Client

With the development of wireless access technology and wireless GRID computing, a new type of *user terminal* called a *wireless thin client* will soon be possible. Compared with the normal *user terminal*, the *wireless thin client* has only a user interface and display capability, and of course radio for the real-time communication with the network. All the data processing and computing are done in the network. Such terminals have low power consumption, and can facilitate the mobile user work or entertainment without the constraint of battery life.

7.1.5   RFID Technology [7]

RFID (radio frequency identification) is a promising technology for the networked sensors. RFID Technology is realized by a *tag* and a *reader*, where

- a *tag* consists of a microchip that stores data and is attached to an antenna as shown in Figure 7.2;
- a *tag* contains a unique serial number, and can have other information such as a customer's account number;
- a *tag* sends the stored data only when being interrogated by the *reader*;
- an *active tag* has an on-tag battery;
- a *passive tag* obtains power from the *reader*;
- a *reader* consists of a radio frequency (RF) module, a control unit and a coupling element to interrogate the *tag* as shown in Figure 7.3;
- a *reader* for a *passive tag* creates an electromagnetic field in the *tag* via inductive coupling or propagation coupling;
- a *reader* receives the data sent by an *active tag*;
- a *reader* has a secondary interface to communicate with backend systems;
- a *reader* has a typical ROM value of 64–128 bits, up to 600 bytes.

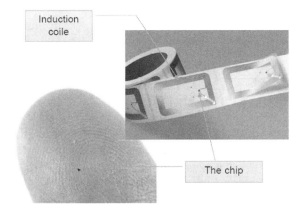

Figure 7.2   An example of an RFID *tag*

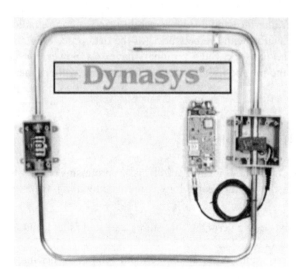

Figure 7.3   An example of an RFID *reader*

Table 7.1   RFID

Frequency band	Characteristics	Typical applications
Low: 100–500 kHz	• Short to medium read range • Inexpensive • Low reading speed	• Access control • Animal identification • Inventory control • Car immobilizer
Intermediate: 10–15 MHz	• Short to medium read range • Potentially inexpensive • Medium reading speed	• Access control • Smart cards
High: 850–950 MHz 2.4–5.8 GHz	• Long read range • Expensive • High reading speed • Requires line of sight (LOS)	• Railroad car monitoring • Toll collection systems

Depending on the application, an RFID device can contain only a *tag*, or only a *reader*, or both a *tag* and a *reader*. The RFID frequency bands, characteristics and typical applications are given in Table 7.1.

RFID at 13.56 MHz band is globally the most used. RFID at 2.4–5.8 GHz may achieve 2 Mbps data rates, with added noise immunity provided by the spread spectrum modulation approach.

Considering that RFID can report what and where the thing is and a sensor can report the condition of that thing, integrated RFID and sensors will certainly have the potential to realize many useful applications in the future, e.g.

- identifying, tracing and observing the status of people and things in real time;
- replacing human senses to monitor the environment.

7.1.6 NFC Technology [7]

NFC (near-field communication) is a technology for very-short-range connectivity with distances measured in centimetres, which is optimized for intuitive, easy and secure connection between various devices without manual configuration by the user.

NFC is based on RFID technology, although with its own frequency band and interface protocols. In short, NFC technology:

- has a range of 0–20 cm, which varies from country to country due to the varying power restrictions;
- operates at 13.59 MHz, which is an unregulated RF band globally;
- communicates in half-duplex, applying a 'listen before you talk' policy;
- uses its own protocols, which

  o distinguish between the *initiator device* and the *target device*, where:

    (a) the *initiator device* initiates and controls the exchange of data,
    (b) the *target device* answers the request from the *initiator device*;

  o distinguish between two operation modes, *active* and *passive mode*, where

    (a) in *active mode* both devices generate their own RF field to send the data;
    (b) in *passive mode* only the *initiator device* generates the RF field while the *target device* uses load modulation to transfer the data.

Initial communication speeds are 106, 212 and 424 kbps, where speed adaptation can be done during the communication on the request of the application and/or environment. NFCIP-1 can use different modulation and bit coding schemes depending on the communication speed, where the *initiator device* starts the communication in a particular mode at a particular speed, and the *target device* determines the current speed and the associated low-level protocol automatically and answers accordingly. Either communication is terminated on command from the application or when the device moves out of the range. NFC-enabled mobile devices are already emerging on the market. NFC-enabled mobile devices utilize xSIM-based security and are capable of communicating with the existing and future electronic world (system, platform, devices), which represents very interesting potential that will play an important role in the NGN world.

7.2   ACCESS NETWORK AREA

Today, an access network is associated not only with a certain type of technology for connectivity, but also with accessible service(s). Under NGN, such primitive dependency between the access technology and service will disappear.

In the NGN environment, an access network provides only the transportation. However, as the first connection for the CTE, the access network remains the most critical part as it is responsible for *ubiquitous connectivity* and *seamless mobility*, both of which are fundamental for *pervasive service accessibility* and *user experience*.

The sophistication of NGN access is that an end-to-end connection can be formed via access networks based on different technologies, and the connection stays as it is when CTEs cross into different radio access technologies.

Where the multiple access networks co-exist, the choice is made according to the access network capability and availability (the presence and the load), terminal capability and status (stationary or on-the-move), service requirement, customer subscription type, etc. Therefore, the access network resources are used in an optimized way. Considering what we have today in the access area, even under overwhelming development of access technologies, especially the wireless access technologies, as shown in Table 7.2, the development space is still tremendous.

7.2.1   Ubiquitous Connectivity

Ubiquitous connectivity in NGN is to enable connectivity for anyone or anything, at any time, from anywhere – in space, on the ground, under the ground, above water, underwater – in any state whether stationary or on-the-move. We classify NGN access development into terrestrial access, space access, underground access, underwater access, sensor access, machine access, ad-hoc access and generic access.

Terrestrial Access

Today, terrestrial access is being mostly developed among other access fields. The available terrestrial access technologies are:

Table 7.2 Existing and emerging access technologies (satellite, mobile, wireless, short-range radio, wired, cable and powerline)

Access type	Technology	Short description
Satellite	UMTS-S	A 3GPP concept, for UMTS satellite components, on-going activity.
	DVB-S	An ETIS concept, for satellite digital video broadcasting, completed. The standard uses MPEG-2 standards for the source coding of audio and video and for the systems level; the latest development of HDTV uses MPEG-4 with AVC (advanced video coding) and more bandwidth-efficient compression.
	DVB-HS	An ETIS concept, for mobile television with satellite digital video broadcasting for handheld devices, on-going activity. The standard uses MPEG-2 standards for the source coding of audio and video and for the systems level; the latest development of HDTV uses MPEG-4 with AVC and more bandwidth-efficient compression.
	DMB (DAB)-S	An ETSI concept of mobile television with satellite digital multimedia broadcasting, completed. It is a further development of DAB (digital multimedia broadcasting). The standard specifies a high spectrum and power-efficient sound and data broadcasting system.
	LEO	An ITU concept, for mobile systems made from LEO (low Earth orbit) satellites.
	MEO	An ITU concept, for mobile systems made from MEO (middle Earth orbit) satellites.
	GEO	An ITU concept, for communication via GEO (geostationary Earth orbit) satellites.
Mobile	IMT- Advanced	An ITU concept, initiated in December 2006, for the IMT-2000 follow-up system. The targeted performance is 1 Gbps bandwidth for the stationary user and hundreds of Mbps bandwidth for the fast-moving user. The targeted commercial deployment is around 2015.
	Next Generation WiMAX	An IEEE concept that will be standardized under IEEE802.16m. The target is to meet the requirement of IMT–Advanced from ITU. The standard is planned to be available in 2009.
	LTE/SAE	A 3GPP concept for UMTS evolution, targeted at smooth evolution to the 4G mobile system by providing 4G performance over the 3G spectrum (>100 Mbps DL and >50 Mbps UL peak data rate, 10 ms end-to-end delay, backwards compatibility to legacy 3GPP access and non-3GPP access, etc.).

Table 7.2 (Continued)

Access type	Technology	Short description
		The LTE will deploy a new air interface OFDMA for DL, and SC-FDMA for UL; its standard is planned to be available in September 2007. The SAE will deploy a new core network with flat architecture, its standard is planned to be available in December 2007. The targeted commercial trial is in 2009, and commercial deployment starting in 2010.
	HSPA	A 3GPP concept for the HSDPA/HSUPA evolution with the consideration of MIMO and a flat architecture core network. It is targeted at optimizing the performance of HSDPA and HSUPA. The relevant standard and targeted commercial trials are in line with LTE/SAE.
	UMTS	A 3GPP concept, the follow-up system of GSM, a 3G mobile system. The available standards are R99 (frozen in 1999), R4 (frozen in 2001), R5 (frozen in 2002), R6 (frozen in 2004) and R7 (frozen in 2007). Achieved 384 kbps DL, 128 kbps UL (WCDMA).
	EDGE	An ETSI concept, evolution of GSM GPRS for mobile data with higher data rate. New coding and modulation schemes have been added to the air interface.
	GPRS	An ETSI concept, evolution of GSM for mobile data. A PS domain has been added to the core network.
	GSM	An ETSI concept of 2G mobile system, a digital system.
	UMB or CDMA2000 Revision C	A 3GPP2 concept for CDMA2000 evolution after Revision B, comparable with UMTS LTE/SAE. The standard is planned for the end of 2007. The targeted commercial trial and deployment will be at the end of 2008.
	CDMA2000 Revision B	A 3GPP2 concept for CDMA2000 evolution after Revision A, comparable with UMTS HSPA. The standard is available.
	CDMA2000 Revision A	A 3GPP2 concept for CDMA2000 evolution after Revision 0, comparable with UMTS HSUPA. The standard is available.
	CDMA2000 Revision 0	A 3GPP2 concept for CDMA2000 evolution after Revision 0, comparable with UMTS HSDPA. The standard is available.

CDMA2000	A 3GPP2 concept for 2G CDMA-one evolution, comparable with UMTS.
UMTS/MBMS	A 3GPP concept for mobile multimedia broadcasting, multicasting and unicasting.
DVB-H	An ETSI concept of mobile television with digital video broadcasting for mobile handheld devices, completed. The standard uses MPEG-2 standards for the source coding of audio and video and for the systems level; the latest development of HDTV uses MPEG-4 with AVC and more bandwidth-efficient compression.
DVB-HS	An ETIS concept, for mobile television with satellite digital video broadcasting for handheld devices, on-going activity. The standard uses MPEG-2 standards for the source coding of audio and video and for the systems level; the latest development of HDTV uses MPEG-4 with AVC and more bandwidth-efficient compression.
DMB (DAB)-T	An ETSI concept of mobile television with terrestrial digital multimedia broadcasting, completed. It is a further development of DAB. The standard specifies a high spectrum and power-efficient sound and data broadcasting system.
MediaFLO	A proprietary system of Qualcomm for mobile television.
ISDB-T	A Japanese digital television system for both mobile and terrestrial television.
IEEE802.16e WiMAX	An IEEE and WiMAX forum concept for mobile broadband Internet access of the Metropolitan Wireless network. The supported full mobility is 120 kmph and the smaller mobility is 60 kmph. The targeted wave 1 products are for embedded laptops coming onto the market. The targeted wave 2 products are for mobile handhelds coming onto the market from 2008.
IEEE802.16m NG WiMAX	An IEEE concept for mobile WMAN, on-going activity. The standard covers advanced the air interface.
IEEE802.20	An IEEE concept for MBWA (mobile broadband wireless access), on-going activity. The standard covers the PHY and MAC layers, operating in the licensed band below 3.5 GHz; optimized for IP data transport; peak data rate per user 1 Mbps; supports mobility up to 250 kmph. Targeted at significantly higher spectrum efficiency, sustained user data rate and capacity than all existing mobile systems.

Table 7.2 (Continued)

Access type	Technology	Short description
	Flash-OFDM	A proprietary system of Qualcomm for mobile broadband access.
	TETRA	An ETSI concept of terrestrial trunked radio for the professional mobile wireless system, completed. A major radio system for the purpose of public safety and security, especially designed for emergency situations and other public safety requirements with faster call set-up times of 300–500 ms (2–3 s for the mobile network).
Wireless	IEEE802.11	An IEEE concept for WLAN, completed in 1997. The standard covers PHY and MAC layers, at 2.4–2.48 GHz ISM band, throughput up to 2 Mbps.
	IEEE802.11a	An IEEE concept for WLAN, completed in 1999. The standard covers PHY and MAC layers, OFDM modulation, at 5.15–5.72 GHz ISM band, throughput up to 54 Mbps.
	IEEE802.11b	An IEEE concept for WLAN, completed in 1999. The standard covers PHY and MAC layers, CCK modulation, at 2.4–2.48 GHz ISM band, throughput up to 11 Mbps.
	IEEE802.11g	An IEEE concept for WLAN, completed in 2003. The standard covers PHY and MAC layers, OFDM modulation, at 2.4–2.48 GHz ISM band, throughput up to 54 Mbps, backwards compatible with 802.11b.
	IEEE802.11n	An IEEE concept for WLAN, on-going activity. Targeted at the enhancements up to five times the throughput and up to twice the range of previous generation WLANs.
	IEEE802.11s	An IEEE concept for WLAN, on-going activity. Targeted at extended service sets – mesh with a wireless distribution system supporting both broadcast/multicast and unicast delivery over self-configuring multihop topologies.
	IEEE802.11y	An IEEE concept for WLAN, on-going activity. Targeted at 3650–3700 MHz operation in the USA.
	IEEE802.16	It is an IEEE concept for WMAN (wireless metropolitan area network), completed in 2002. The standard covers the PHY and MAC layers, at bands between 2 and 66 GHz with channel widths between 1.25 and 20 MHz; providing LOS fixed wireless access and point-to-multipoint links for reliable, high-speed network access in the first mile and last mile at home or in an enterprise.

IEEE802.16a	An IEEE concept for WMAN, completed in 2003. The standard extends 802.16 to licensed and licence-exempt bands from 2 to 11 GHz.
IEEE802.16d	An IEEE concept for WMAN, completed in 2004. The standard extends 802.16 to the LOS solution over 10.5, 25, 26, 31, 38 and 39 GHz; and to the NLOS solution over both license and licence-exempt frequencies.
IEEE802.16e	An IEEE concept for mobile WMAN, completed in 2005. The standard covers PHY and MAC layer to support mobility for frequency <6 GHz.
IEEE802.16j	An IEEE concept for mobile WMAN, on-going activity. The standard covers multihop relay specification.
IEEE802.22	An IEEE concept for WRANs (wireless regional area networks), on-going activity. The standard covers PHY and MAC layers, a cognitive radio-based spectrum air interface for use by licence-exempt devices on a non-interfering basis in spectrum allocated to the television broadcasting service.
HIPERLAN /1	An ETSI concept for radio LAN (radio local area network), completed. The standard is designed for high-speed communications (20 Mbps) between portable devices at 5 GHz frequency range; intended to allow the flexible wireless data networks to be created without the need for an existing wired infrastructure; can be used as an extension for a wired LAN.
BRAN HIPERLAN /2	An ETSI concept for the short-range variant of BRAN (broadband radio access network), completed. The standard is designed for complementary access mechanisms for UMTS system and for private use as a wireless LAN; offers high-speed access up to 54 Mbps at 5 GHz frequency range.
BRAN HIPER-ACCESS	An ETSI concept for the long-range variant of BRAN, completed. The standard is intended for point-to-multipoint, high-speed access with 25 Mbps as the typical data rate by residential and small business users to a variety of networks; the spectrum allocation is 40.5–43.5 GHz.
BRAN HIPERLINK	An ETSI concept for another variant of BRAN, completed. The standard is intended for short-range very-high-speed interconnection between HIPERLAN and HIPERACCES, e.g. up to 155 Mbps over distances of up to 150 m; the spectrum is in the 17 GHz range.
DECT	An ETSI concept for a high-capacity digital radio access technology for cordless communications in residential, business and public environments. It is also one of the five air interfaces from IMT-2000 of ITU, completed.

Table 7.2 (Continued)

Access type	Technology	Short description
		The standard is designed for short-range use as an access mechanism to the main networks; it offers cordless voice, fax, high-speed data and multimedia communications, wireless local area networks and wireless PBX; at frequencies of 1880–1990 MHz.
	DVB-T	An ETSI concept for terrestrial digital video broadcasting, completed. The standard uses MPEG-2 standards for the source coding of audio and video and for the systems level; the latest development of HDTV uses MPEG-4 with AVC and more bandwidth-efficient compression.
Short-range radio	IEEE802.15 Bluetooth	An IEEE concept for WPAN, completed in 2002. The standard is for the PHY and MAC layers, operating at 2.4 GHz ISM band for 10 m coverage.
	IEEE802.15.3a HRD UWB	An IEEE concept for WPAN of HRD UWB (high-rate data UWB), completed in 2004. The standard covers the PHY and MAC layers for HRD UWB providing data rates of 20 Mbps to 1 Gbps; range 1–20 m, over the 3.1–1.6 GHz band.
	IEEE802.15.3c	An IEEE concept for WPAN, on-going activity. The standard covers a millimetre-wave-based alternative PHY layer; operating in new and clear bands including the 57.64 GHz FCC unlicensed band; targeted at very high data rates; optional data rate >2 Gbps; high coexistence with all other microwave WPANs.
	IEEE802.15.4a LRD UWB	An IEEE concept for WPAN of LRD UWB (low rate data UWB), completed. The standard covers the PHY and MAC layers for LRD UWB providing data rates of around 1 Mbps; range 50 m.
	IEEE802.15.4 ZigBee	An IEEE concept for WPAN, completed. It is a set of high-level protocols, operating on top of IEEE802.15.4 PHY and MAC layers. The IEEE802.15.4 PHY and MAC layers are designed for small, low-complexity, low-power PAN and sensor applications; operating in the 2.4 GHz, 915 MHz and 868 MHz ISM bands; providing data rates of 20–25 kbps; range 10–30 m with 0.5–1 mW EIRP.
	IEEE802.15.4d	An IEEE concept for WPAN, on-going activity. It is for the alternative PHY layer extension to support the Japanese 950 MHz band.
	IEEE802.15.5	An IEEE concept for recommended practices for mesh topology capability in WPAN, on-going activity. The standard covers the PHY and MAC layers for WPANs to enable mesh networking.

	Wibree	An Open Industry Initiative led by Nokia, on-going activity. Targeted at connecting small, very-low-power devices such as watches and medical and sports sensors to larger equipment such as mobile phones and PCs; up to 1 Mbps over distances of 5–10 m. The standard is planned for the H2 of 2007.
	NFC	A concept of NFC (near-field communication) forum, on-going activity. It is based on RFID, very short access range of 0–20 cm; providing 424 kbps data rate; operating at 13.56 MHz band and over its own protocols. It is optimized for intuitive, easy and secure connection between various devices without manual configuration by the user.
	Infrared IrDA	A concept of IrDA (infra red data association), completed. It uses infrared light at 850–900 nm, and covers the range up to 100 cm.
Wireline	IEEE802.3ah EPON	An IEEE concept for 1 Gbps ethernet PON (passive optical networks), completed. It provides 1 Gbps capacity with unlimited maximum logical reach and range.
	IEEE802.3av EPON	An IEEE concept for 10 Gbps ethernet PON, on-going activity. It provides 10 Gbps capacity with symmetric and asymmetric configurations, and unlimited maximum logical reach and range.
	ITU-T G.984 GPON	An ITU concept for Gigabit-Capable PON, completed. It has limited maximum logical reach (60 km) and range (20 km).
	xDSL	An ITU concept, including ISDN, ADSL, HDSL and VDSL. It is designed to operate on telephone wires intended originally for voice-band communication (300 Hz to 3.4 kHz), bandwidth varying from narrow-band ISDN <100 kHz to VDSL >10 MHz.
	FTTx	An ITU concept including FTTH, FTTB, FTTC, FTTP etc.
Cable	DVB-C	An ETSI concept for cable digital video broadcasting, completed. The standard uses MPEG-2 standards for the source coding of audio and video and for the systems level; the latest development of HDTV uses MPEG-4 with AVC and more bandwidth-efficient compression.
Powerline	PLC	A concept of OPERA (Open PLC European Research Alliance) for PLC. The specification has been submitted to both the IEEE and ETSI. Targeted at a low-cost broadband over electricity networks for high-speed Internet access, VoIP telephony and video-on-demand, etc.

- wired access technologies such as

  ○ FTTx (fibre to the home, fibre to the building, etc.),
  ○ xDSL (digital subscription line),
  ○ PLC (power line communication),
  ○ broadcasting cable;

- fixed wireless access technologies such as

  ○ fixed WiMAX (wireless metropolitan),
  ○ HIPERAccess,
  ○ DECT (digital enhanced cordless telecommunications, former for digital European cordless telephony),
  ○ LMDS (local multipoint distribution system),
  ○ MMDS (multichannel multipoint distribution systems),
  ○ MVDS (multichannel video distribution system);

- nomadic wireless access technologies such as

  ○ DECT,
  ○ WLANs,
  ○ Bluetooth,
  ○ UWB (ultra wide band);

- mobile access technologies such as

  ○ GSM (global system for mobile communications),
  ○ UMTS (universal mobile telecommunications system),
  ○ CDMA2000,
  ○ mobile WiMAX (IEEE802.16e),
  ○ TETRA (terrestrial trunked radio);

- wireless broadcasting technologies such as

  ○ DAB (digital audio broadcasting),
  ○ DMB-T (terrestrial digital multimedia broadcasting),
  ○ DVB-T (terrestrial digital video broadcasting),
  ○ DVB-H (handheld digital video broadcasting),
  ○ ISDB-T,
  ○ MediaFLO,
  ○ MBMS (multiple broadcast multimedia services).

However, the wired access technologies are for broadband, and the mobile access technologies are for mobility. With the upcoming *wireless thin client-*type terminals and three-dimensional mobile multimedia gaming services, the demand on the access technology will be for extremely high bandwidths while terminals are moving at high speeds. Today trains can reach speeds of 580 kmph, faster than the take-off speed of an airplane!

The on-going developments for next-generation mobile access technologies, targeted at 100s Mbps bandwidth and for 500 kmph moving speed, are:

- NGMN (next generation mobile network) from NGMN Alliance (formerly NGMN initiatives [1]);
- LTE (long-term evolution) from 3GPP [2];
- UMB (ultra mobile broadband) or CDMA2000 Rev. C from 3GPP2 [3];
- next generation WiMAX IEEE 802.16m from IEEE [4];
- IMT-Advanced from ITU [5].

EU research programmes FP6 and FP7 are also very active in this development field [6].

As of June 2007, the latest achievement at NTT DoCoMo was 5 Gbps access speed for a stationary terminal using 12×12 MIMO antenna and OFDMA technologies.

Space Access and Remote Terrestrial Access

Today, space access via satellite is already possible, although in narrow band only. Space access is used for communication from spaceships and airplanes, and also for remote terrestrial areas, e.g. in deeply forested areas, at sea or on mountains. Existing technologies or on-going developments include:

- satellite communication systems of GEO, MEO and LEO;
- the satellite component of UMTS UMTS-S;
- the satellite component of digital video broadcasting, DVB-S.

However, broadband space access is needed for the real-time multimedia monitoring of flying objects like airplanes, and for DSL-like connectivity for end-users within a large airplane in flight. Development will be focused on broadband space access with tens of Mbps bandwidth for moving speeds above 1000 kmph.

Underground Access

In the NGN environment, underground access is responsible for connectivity inside a tunnel or deep in the earth. What is available today is repeater technology for access underground or inside tunnels, which is very limited in terms of availability and quality. Development will be focused on access technology in which the earth can be seen as a transparent layer to a certain extent.

Underwater Access

In the NGN environment, underwater access is needed for monitoring life deep
in the sea and the movement of the seabed, besides underwater communications.
What is available today is far from adequate. Development will be devoted to
access technology in which the water is seen as a transparent layer to a certain
extent.

Sensor Access

In an NGN environment, sensors are embedded everywhere, especially in those
places where human beings have difficulty in gaining access, e.g. in space,
underground and deep in the sea, in order to collect data from nature and to
be aware of changes, e.g. sea pollution, or to predict natural disasters like
earthquakes or seaquakes.

   As the location of a sensor can change if it moves with the seabed, and
sensory information is sent only when unusual values have been measured due
to the power consumption, the access technologies for networking sensors have
to be very reliable and ubiquitous.

Generic Access

Today, communications between terminals are almost always via core networks
even when they are located next to each other. Owing to the scarcity of radio
resources, we will not be able to afford such a luxury in the future. NGN
should also facilitate the realization of the generic access concept, as shown in
Figure 7.4, in which:

- with an integrated radio, everything can become a network node as a router
  or a repeater, e.g. a table, a chair, a bookshelf, a tea pot, a shoe, a bag, a car;
- such a network node from anything can be connected to everything in an
  ad-hoc way;
- the connection between such network nodes can be based on any radio
  technology, e.g. WLAN, WiMAX, 802.20, RFID, ZigBee, UWB.

   In order to realize generic access of NGN, developments are needed:

- very small radio elements that can be stuck to anything without affecting
  its appearance or even appearing as a decoration;
- standardized and robust routing protocols.

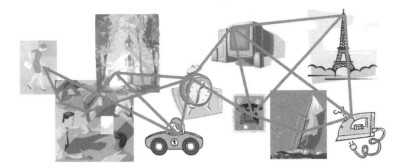

Figure 7.4   A future generic access scenario

Seamless Mobility

NGN access network is also responsible for *seamless mobility*, i.e.

- supporting seamless connectivity for terminals on-the-move, crossing the borders of a cell or network;
- supporting seamless connectivity for terminals crossing heterogeneous radio access networks among satellite, mobile, fixed wireless and short-range radio;
- supporting extremely fast-moving terminals with high performance, which is essential for real-time video monitoring on airplanes or spacecraft.

To satisfy these requirements, the handover or roaming mechanisms of today are far from adequate and further developments are needed.

Quality of Access Connectivity

As the bottleneck, the access network is usually responsible for the poor quality of an end-to-end connectivity, not only owing to the bandwidth but also owing to the latency of the access generated. To reduce the latency of the access network is another development dimension.

7.2.2   Co-existence Mechanisms for Multiple Radio Access Networks

No doubt, the overwhelming development of radio access technologies including satellite, mobile and wireless will benefit NGN's ubiquitous accessibility, but it could also impossibly complicate radio design for their co-existence without strict frequency regulation today. However, owing to the extremely low usage of radio resources (only 5 % usage on average), frequency re-farming has been proposed. The trend is to go for technology-neutral spectrum assignment, which means co-bands for different radio technologies, which could lead to interference being disruptive for performance.

Furthermore, the new radio technologies like UWB utilize a band from 3.1 GHz to 10.7 GHz. The huge bandwidth of UWB (7.6 GHz) leads to the necessary co-use of spectrum between UWB and the existing and emerging wireless systems. UWB-like access technologies represent the requirement for novel spectrum management. Therefore, intelligent mechanisms are needed for the co-existence of multiple access technologies. The adaptive radio technologies are promising; however they are still in the early stages of development:

- *Software radio* is an interface that can, if necessary, change within a few packets to a completely different interface in a different band that can be another radio access technology or another band of the same radio technology. It uses agents in the terminal (e.g. a software channel booking agent) to negotiate with the agents in the network to obtain the best deal (service, QoS, price) for the user-requested service.
- *Cognitive radio* is similar to software radio in that it is highly adaptive to user needs and the actual radio environment (noise, interference, etc.) and is able to learn and adapt automatically.
- *Software-defined radio* (SDR) is much more advanced than software radio and cognitive radio. SDR systems (including terminals and end-to-end network nodes) are re-configurable, allowing flexible spectrum allocation and seamless inter-working between technologies. The agents sit at a terminal and network nodes, and re-configure and control the terminal and those network nodes for the radio access technology, frequency band, modulation type and output power according to the user's needs, radio environment, radio resource availability, terminal capability, network infrastructure capabilities, requested service and traffic distribution.

7.3 BACKHAUL NETWORK AREA

Today, the most used backhauls for mobile or wireless radio access network are E1 and T1. The concept of next-generation radio base stations includes multi-technologies, multibands, VDSL-like broadband, flexible capacity scalability and solar-powered, small, light and easy to install devices.

FTTx is of course such a backhaul solution, however a novel wireless backhaul solution is also needed for rapid deployment of a mobile network or for those areas where optical connection is not available. The wireless mesh network is a promising solution, as shown in Figure 7.5, where other Node Bs relay the backhaul for most of the Node Bs before being connected to the core network. The current wireless mesh technology is still far from the expected carrier-grade backhaul for large-scale deployment, predictable

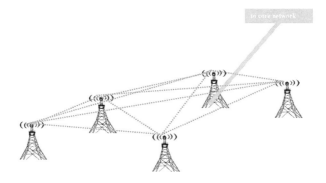

Figure 7.5   A wireless mesh network backhaul

performance and the built-in mechanisms for SON (self-organizing network) covering self-configuration, self-tuning and self-healing. The technology also needs to work on the topology, protocols and algorithms to overcome the weaknesses of wireless mesh network of today, such as:

- rapid performance (throughput and latency) degradation with the increased hop number;
- poor scalability;
- less suitable to simultaneously support mixed services from voice, video and data.

The on-going WLAN-based mesh network (IEEE802.11s) is working on such a solution.

7.4   CORE TRANSPORT NETWORK AREA [9]

For many years, the core transport networks have been in transformation towards an all-IP-based packet transport network. The transformation started from the backbone and has extended stepwise to the network edge.

The on-going transformation for the core transport network is from the best-effort IP network towards the managed IP network, which will guarantee that sufficient QoS is delivered with the customer services. There already exist many IP QoS supporting protocols, e.g. RSVP, IntServ and IntServ/RSVP, DiffServ, MPLS and DiffServ/MPLS, which are described briefly in Table 7.3.

However, all these have problems in providing a satisfactory solution, e.g.

- the IntServ/RSVP provides guaranteed end-to-end QoS, but is too complex and not scalable, and thus is only suitable for small networks;
- the MPLS and DiffServ provide only relative QoS (better than best-effort), but are scalable.

Table 7.3 IP QoS supporting protocols

Protocol name	Short description
RSVP	RSVP (resource reservation protocol), defined by IETF RFC 2205, is a signalling protocol for the network resource reservation. RSVP offers a mechanism for signalling QoS across the network. By provisioning resources for network traffic, RSVP can assist packet switching protocol, e.g. IntServ or DiffServ, to provide service discrimination according to the QoS class, e.g. to allocate network resources explicitly for delay sensitive applications. RSVP is an out-of-band signalling system. RSVP messages are sent e2e, but carry a flag to enable them to be read and processed by network elements. RSVP is a two-path protocol, able to handle asymmetrical paths. RSVP assumes that the receiver is responsible for establishing QoS (paying for the level of QoS it receives). The sender initially describes the data and identifies the data path; the receiver sends a reservation message back along the data path. In detail, the RSVP works as described below: (a) The sender generates so-called PATH message, which propagates through the network to the receiver, gathering information about the network. The message contents consist of a number of objects including the session identifier, the previous (RSVP-enabled) hop address, the maximum transmission unit size, the minimum path latency (a distance-related transmission delay), the path bandwidth estimation (the amount of bandwidth the receiver could ask for within the service, which bears no relationship to the bandwidth the sender might want) and the parameters that enable the receiver to calculate routing delay. (b) The receiver sends a so-called RESV message, which chooses the required QoS and establishes the reservation. The RESV message is propagated back along the same path through the network via each of the previous router addresses, as stored during the PATH stage (note: unlike within Internet, messages flowing in opposite directions between two terminals will usually follow different paths). (c) Both sender and receiver can generate messages to delete the reservation, the PATH_TEAR message by the sender and the RESV_TEAR message by the receiver. RSVP has been widely implemented on a range of different terminals, including Microsoft Windows. RSVP is a key element of IntServ which is described later.

IntServ   IntServ (integrated services), defined by IETF RFC 1633, is a packet-switching protocol. IntServ is designed to support integrated services – the transport of audio, video, real-time and classic data traffic within a single network infrastructure.

DiffServ   DiffServ (differentiated services), defined by IETF RFC 2475, is a packet-switching protocol. DiffServ provides a coarse and simple way to categorize and prioritize network traffic (flow) aggregates with no signalling mechanism. The DiffServ-defined QoS classes are:

- one expedited forwarding (EF) class – aims to provide a low-jitter, low delay service for traffic;
- four assured forwarding (AF) classes – aim to provide delay-tolerant applications at three levels for each class, based on the drop precedence;
- best-effort class.

DiffServ provides relative QoS, which means that one is handled better than the other, enabling a 'better than best-effort services' over the IP network domain.

DiffServ-provided QoS is scalable, because the routers in the backbone network do not need to keep information of the individual flow; they do their job simply by examining a single field of the IP header.

However, DiffServ provides no absolute QoS guarantees with regard to bandwidth and delay for a traffic flow. In principle, appropriate network dimensioning, admission control and traffic engineering can overcome the non-guaranteed QoS. The DiffServ QoS is only delivered to aggregated traffic classes (rather than specific flows).

The basic idea of DiffServ is that the IP header of each packet within a user session (flow) will be marked with a special code, DSCP (DiffServ code point), at the edge of the network, indicating the relative priority of the packet. The DiffServ-enabled routers will schedule the packets based on the DSCP value. The DiffServ functions operate on the aggregated flows and perform shaping and dropping to bring the stream into compliance with the requested QoS. In detail, as shown in Figure 7.6:

(a) On entry to a network, the packets are placed into a broad service group by the classifier.
(b) The classifier reads the DSCP in the IP packet header as shown in Figures 7.7 and 7.8, the source and the destination address, and determines the correct service group (the correct group or class is determined through static SLA for the prioritization).

Table 7.3 (Continued)

Protocol name	Short description
	(c) The packets are then given a suitable marking, which may involve changing the DSCP.
	(d) Traffic shaping may then occur to prevent large clumps of data with the same DSCP entering the network. All packets within a specific group receive the same scheduling behaviour. These behaviours can be simple to implement using class-based queuing. The shaper may delay some or all of the packets in the traffic stream in order to bring the stream into compliance with the requested traffic profile. In the downlink direction, shaping is a crucial function. GGSN will shape each downlink data flow (i.e. a GTP tunnel). This shaper operates on the GTP tunnel (i.e. on the individual user flow to one user), thus the radio resources are protected from being overloaded by long bursts from a single user. It protects the system from denial of service attacks and allows for fair sharing of the scarce radio resources. The shaper may also discard packets in the traffic stream to handle congestion. In the case of congestion, an early drop algorithm will drop packets to minimize the disturbance seen from the end-user. Packets will also be dropped if the shaper overflows.
	(e) Once in the network, routers only have to forward the packets according to these network-defined scheduling behaviours, as identified through the DSCP. The complex processes 'classification, marking, policing' to ensure that no class is over-subscribed, and 'traffic shaping' only takes place at the boundaries of each network domain. This may be done individually by the traffic sources or edge nodes, or even a centralized bandwidth broker may be involved. This is sufficient to protect the network and guarantee the service for the aggregate class.
	DiffServ appears to be a good solution to part of the QoS problem as it removes the per-flow state and scheduling that lead to scalability problems within the IntServ architecture. However, it provides only a static QoS configuration, typically through service level agreements, as there is no signalling for the negotiation of QoS. As with MPLS, e2e QoS cannot be guaranteed. DiffServ was only ever intended to be a scalable part of an e2e QoS solution.
MPLS	MPLS (multiprotocol label switching), defined by IETF, is a packet-switching protocol.
	MPLS was designed originally to improve the forwarding speed of routers. However MPLS also enables QoS differentiation, particularly suited to carrying IP traffic over fast ATM networks.

MPLS can provide a wide range of differentiated QoS classes, which are guaranteed only across the MPLS domain, not on an e2e basis.

The MPLS-enabled differentiated QoS classes cover the reliable data transport and delay-sensitive transport services. SLAs are typically used for admission control for prioritizing services.

The basic idea of MPLS is to add labels in packet headers and provide bandwidth management for aggregates via network routing control. MPLS uses marking (20 bit labels) to determine the next router hop. In detail:

(a) Routers at the edge of MPLS domain mark all packets with a fixed-length label (20 bits).
(b) The label acts as shorthand for the information contained in the IP packet header. MPLS resides only on routers.

The MPLS label identifies both the route through the network and the QoS class of the packet, therefore MPLS packets can follow the pre-determined paths according to the traffic engineering and specified QoS levels.

As the MPLS label is very short (20 bit), packets can be routed very quickly, requiring significantly less processing than routing based on IP header.

MPLS combines layer 2 switching technology with layer 3 network layer services and therefore reduces the complexity and operation costs.

Depending on operator-selected transport technology, the QoS control access to the IP backbone of CN may be supported by MPLS. For scalability reason, this QoS control will be done for traffic aggregates. All packets belonging to the same class will then be treated in the same way in the network, which means that they will be put in the same forwarding queues in the routers of the network.

In many respects, MPLS for QoS is similar to the DiffServ approach, although, to reduce the scalability problems, it is usually used as a layer 2 rather than a layer 3 solution.

MPLS provides improved granularity of service at the expense of more complex administration. In itself, it cannot provide e2e QoS configurable on a flow-by-flow basis.

Table 7.3 (Continued)

Protocol name	Short description
IntServ/RSVP	IntServ/RSVP uses RSVP to provide the reservation set-up and control to enable the IntServ, the RSVP reservation, to be regularly refreshed. IntServ/RSVP can provide differentiated QoS over an e2e path in IP domain through bandwidth reservation, which is per-flow-based and for aggregates. The IntServ/RSVP architecture can provide three basic e2e QoS classes: • Guaranteed service, which offers hard QoS, guaranteed for (a) quantified delay and jitter bounds and (b) no packet loss from data buffers (ensuring near loss-less transmission), is intended to support real-time traffic;. • Controlled load service, which makes the network service appear to be a lightly loaded best-effort network, aims at delay-tolerant applications. • Best-effort, which requires no reservation. The IntServ/RSVP provides hard guaranteed QoS. However, the IntServ/RSVP approach provides poor scalability and is only suitable for small networks. IntServ/RSVP duplicates some of functionality already provided in RTP (real time protocols), such as jitter control. Compared with DiffServ, the IntServ/RSVP is more complex and more demanding for application hosts and for network elements (routers and switches).
MPLS/DiffServ	DiffServ/MPLS, a DiffServ-based MPLS network, remains to ensure scalability with both DiffServ and MPLS, driving complexity towards the extremes while providing better QoS assurance through path reservations over a LSP (label switched path).

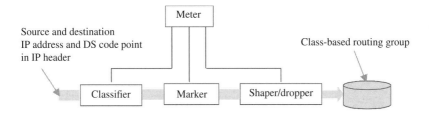

Figure 7.6   Component in a DiffServ border router

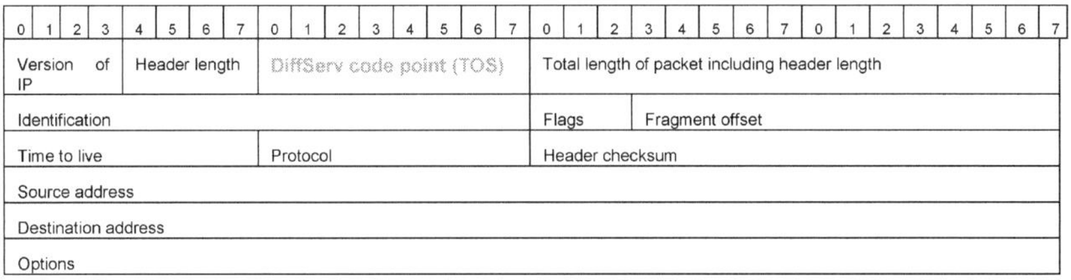

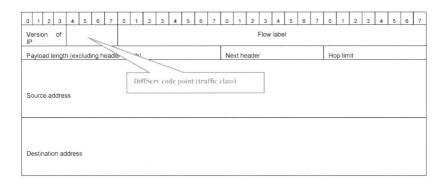

Figure 7.7   IPv4 packet header format; the DiffServ code point is in the TOS (type of service) field

Figure 7.8   IPv6 packet header format; the DiffServ code point is in the traffic class field

The problem is about end-to-end QoS: those protocols can only provide QoS within their own domains. The key development in core transport area will be a unified network layer protocol supporting the end-to-end QoS with fine granularity, which can be understood and interpreted by all the underlying physical networks, ATM or PSTN, etc., which are interworking or interconnected.

7.5   SERVICE CREATION AREA

Today, service providers are still desperately searching for the killer application in order to pay back their network investment, for example, video telephony was once hoped to return the investment in the mobile 3G network. The reality that operators face is that the comfortable time has passed where they can live on a single service or a dominant service in the lifetime of a network. This reality requires another kind of business scenario that is based on extremely rich offerings of services, applications, content and information whose lifetimes are limited or even very short.

The NGN promises prosperous services for the end-user and the freedom for each end-user to become a service provider. This demands an advanced service creation environment. Such an advanced service creation environment should facilitate a standardized way to develop service, which is network-independent and therefore is for converged service development.

As fundamentals for NGN, the essential intelligent services included in this environment are:

- on-line translation for real-time communication in different languages, including sign language for hearing- or speech-impaired people;
- semantic searching according to concepts instead of key words, e.g. to search for a holiday destination by giving a semantic information like 'sunny, German-speaking, 2000 Euro budget amount, May 1–20 2007'.

In the NGN architecture, this advanced service creation environment includes the *service support functions* and the *open interface* towards the *service/ application/content/information layer*, where:

- The *enablers of non-network services or network services compose the service support functions*. For network services, the enablers are responsible for communicating with the underlying networks. The typical enablers are:
  - ○ real-time language translation;
  - ○ semantic searching;
  - ○ content processing;
  - ○ call control;
  - ○ session control;
  - ○ messaging;
  - ○ location information;
  - ○ presence information;
  - ○ group communication management;
  - ○ broadcasting/multicasting management;
  - ○ number portability management;
  - ○ data synchronization.

- The *Open Interface* is a standardized language for applications to call those basic network services via enablers to build up more advanced services. Such languages are called application programming interfaces, APIs.

Various efforts have already been made and will continue to realize the above-described service creation environment, e.g. OSA/Parlay and Parlay X. However, owing to the unnecessary fear of operators being degraded to pipelines of bits, the progress has been very slow for the development of such a future-oriented Service creation environment. The actual situation is that the enablers have never been seriously placed and the APIs used are still proprietary and primitive at operators.

7.5.1 OSA/Parlay Technologies [12, 13]

OSA/Parlay technology defines a standardized, secured and controllable set of APIs for any trusted or non-trusted service provider to access the basic network services to build up advanced services. The OSA architecture is given in Figure 7.9, in which:

- the OSA gateway is a layer between the network and applications;
- the OSA gateway is composed by the FW (framework) and the SCSs (service capability servers), which are not necessarily co-located geographically;
- OSA API is the language for application servers communicating with an OSA gateway to call an SCF (service capability feature) sited at the SCS.
- the OSA gateway communicates with network elements through standardized or proprietary protocols.

The FW provides management functions to applications, including:

- authentication – the application and OSA gateway authenticate each other bi-directionally;
- authorization – the application is authorized to use only agreed SCFs;
- discovery of SCFs – applications learn available SCFs;
- registration of new SCFs – new SCFs are registered to FW;
- establishment of service agreement – SLA establishment between applications and gateway for the usage of SCFs
- access to SCFs – applications access the agreed SCFs;
- operational management functionality, initial access, event notification, heartbeat management, fault management, load management, OAM, etc.

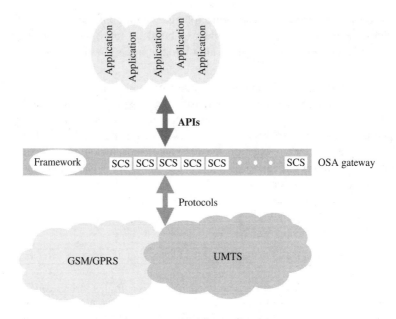

Figure 7.9   OSA/Parlay architecture

The SCSs provide the following SCFs to applications:

- call control for generic call control, multiparty call control, multimedia call control and conference call control – to set up basic call and manipulating multimedia and multiparty conference calls;
- user interaction – to provide voice or data exchange between the application and the end-user;
- mobility for user status, user location, user location CAMEL and user location emergency – to provide the location and status of a single user or multiple users;
- terminal capability – to provide the terminal capabilities of an end-user;
- Data session control – to provide data session control for the IP network;
- Generic messaging – to provide access to SMS mailboxes, multimedia messaging, etc.;
- Connectivity management – to provide provisioned QoS;
- Account management – to provide access to the accounts of an end-user;
- Charging – to provide on-line charging and off-line charging to the end-user for the usage of applications/data;
- Policy management – to provide policy-enabled services;
- PAM (presence and availability, event management, provisioning) – to cooperate with network services like call control and user location information, and the non-network services like user account information and content-based charging.

7.5.2 Parlay X Technology

Parlay X technology, a new effort after OSA/Parlay, was initiated to specify a new set of APIs called Parlay X APIs:

- Parlay X APIs are the light version of OSA APIs, powerful yet simple, based on the principle that 80 % of OSA applications use only 20 % of OSA functions;
- Parlay X APIs can be realized using a simple XML-based message exchange.

7.5.3 Web 2.0

Web 2.0 represents a fundamental advance of the Internet, which enables any end-user to customize his or her own environment and to become a provider of services, applications, content and information.

7.6 NETWORK CONTROL AND MANAGEMENT AREA [8]

Today, the network control and the network management are separated. The network control is for call set-up, call tear-down, etc. and network management concerns the management of configuration, faults, performance, alarms and inventory. In the NGN environment, the network control is constantly present, not only at call-starting and call-ending. The network control is constantly in action to deliver performance at the right level. This constant control is based on the information provided by the management functions, e.g. the monitored performance provided by the performance management.

Therefore, there is a common area in NGN architecture called the *network control and management area*, as indicated in Figure 7.1. This area is responsible for controlling and managing end-to-end connectivity across various types of access networks and heterogeneous core transport networks according to:

- the network capability, the network availability, the network performance and the network provider;
- the requested service, i.e. the service priority, the service required QoS and the service provider;
- the served customer, i.e. the customer subscription, the actual customer SLA satisfaction level;
- the user terminal, i.e. the terminal capability, the terminal status (stationary, on-the-move, speed of the move, etc.).

The control and management of an end-to-end connectivity includes:

- setting up, maintaining and tearing down the end-to-end connectivity;
- monitoring the performance of end-to-end connectivity and initiating action when necessary;
- monitoring the available access networks, the load on them and their performance;
- managing the addition of new access networks, upgrading existing ones and removing old ones;
- monitoring the available core transport networks, the load on them and their performance;
- managing the addition of new transport core networks, upgrading existing ones and removing old ones;
- monitoring terminal status;
- analysing the performance of end-to-end transport chains, predicting problems and activating the relevant measure;
- detecting problems in end-to-end transport chains, analysing the consequences and organizing the relevant information to the relevant manager or team;
- providing billing information according to the traffic volume.

In an ideal case, this area should function as described in the following subsections.

7.6.1   Setting up, Maintaining and Tearing Down End-to-End Connectivity

Before Terminal A is Switched On

The *network control and management* monitors the access networks and the core networks constantly (Figure 7.10).

Terminal A is Switched On (Idle State)

Terminal A is connected to the network via a master access network and reports to the *network control and management* about its capability, status and the detected access networks (Figure 7.11).

Terminal A Requests a Service (Active State)

The *network control and management* consults the *service control and management* about the following:

- service priority;
- connection type (PtP, PtMP, MPtMP, simplex or duplex);
- QoS requirements (Figure 7.12).

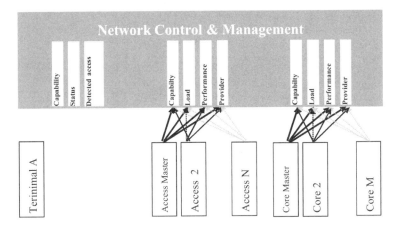

Figure 7.10   Before terminal A is switched on

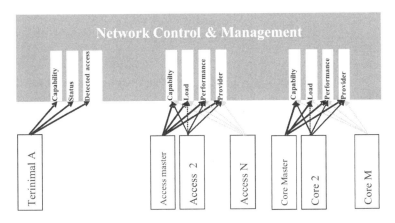

Figure 7.11   NGN *network control and management*

Setting up end-to-end Connectivity for Terminal A

Taking into account all the factors, the *network control and management* will set up end-to-end connectivity for terminal A via signalling, including:

- which access network to take;
- which core transport network to take;
- what kind of performance for the end-to-end connectivity it should be.

Maintaining end-to-end Connectivity for Terminal A

The performance of end-to-end connectivity is constantly monitored. When the connectivity is under-performing, the network control and management will terminate the connection and a new end-to-end connection will be established.

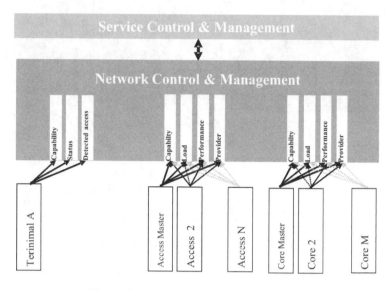

Figure 7.12   Terminal A requests a service

Tear Down the Connection when the Requested Service is Terminated for
Terminal A

   When the service is successfully terminated, the *network control and
management* will tear down the connection, and all the network resources are
released.

7.6.2   Monitoring and Controlling the Performance of End-to-End Connectivity

The performance of end-to-end connectivity is constantly monitored. When
the connection is under-performing, the *network control and management* will
allocate the problem and activate the relevant measures, e.g. changing the
access network or changing the core transport network.

7.6.3   Analysing and Predicting Performance of End-to-End Connectivity

The end-to-end connectivity performance is constantly analysed. When a
possible problem is predicted, the *network control and management* will activate a relevant measure to prevent it occurring.

7.6.4 Generating and Delivering Relevant Information to the Relevant People

When a problem or problems are detected along the end-to-end transport chain, the consequences will be analysed and then organized into the relevant information to be delivered to the relevant management and operation teams.

7.6.5 Generating Billing Information

After tearing down the connection, the *network control and management* will generate charging information and the SLA-related information and report to the *service control and management*.

7.6.6 Managing Multiple Access Networks Belonging to Different Operators

Any access network provider can offer its access connectivity service to the *network control and management* under an agreed SLA, or withdraw from the service.

7.6.7 Managing Multiple Core Transport Networks Belonging to Different Operators

Any core transport network provider can offer its transport connectivity service to the *network control and management* under an agreed SLA, or withdraw from the service.

7.6.8 Managing Changes in the Access Network

Any access network provider has the freedom to upgrade the access network and inform the *network control and management* about its new capabilities.

7.6.9 Managing Changes in the Core Transport Network

Any core network provider has the freedom to upgrade the core transport network and inform the *network control and management* about its new capabilities.

7.6.10   End-to-End Network Resource Management

The *network control and management* also allows:

- the addition of a new access network, to upgrade an existing access network and to remove an old access network;
- the addition of a new core transport network, to upgrade an existing core transport network and to remove an old core transport network.

7.7   SERVICE CONTROL AND MANAGEMENT

As mentioned in the previous section, the service management is very poor in today's network. In the NGN environment, owing to the customer-centred operation, service management is a key element. As defined in the NGN architecture shown in Figure 7.1, aspects of *service control and management* cover to be developed or further developed are listed below (Figure 7.13).

Service control and delivery will make available the components of *IMS* for SIP-based services; *PSTN/ISDN emulation* for enabling the traditional PSTN/ISDN services; *streaming* for content delivery, broadcast/multicast and push services; and *other multimedia services* for retrieval applications, data communications services, on-line applications, sensor network services and remote control services.

GRID is a candidate component, which will enable GRID computing over wired or wireless networks.

Business policy management defines the priority level for the delivery of a service according to the business priority of the service, the service provider,

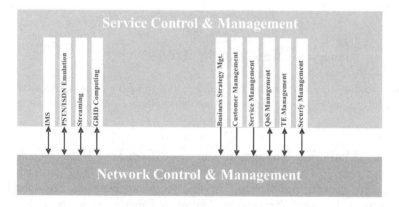

Figure 7.13   NGN *service control and management*

the customer subscription and the customer's actual SLA satisfaction level. The business policy can be changed following a business decision of an operator.

The priority level of a service can vary with:

- the time, e.g. higher priority can be given during a promotional campaign, lower priority during weekends;
- the provider, e.g. higher priority for providers offering better revenue sharing agreements;
- the customer subscription, e.g. customers with gold class subscription have the highest priority for network resources;
- the customer's actual SLA satisfaction level, e.g. those customers whose SLA is closer to being broken are given the highest priority for network resources.

Customer management – in addition to the in-place customer subscription and customer service provisioning, the customer-centred approach requires:

- customer self-care for self-subscription, self-provisioning and self-checking of own account;
- customer account management for account balance, m-commerce, etc.;
- advanced customer billing/charging management by defining the billing/charging policy according to

  ○ the service, service provider, customer, QoS and SLA satisfaction level;
  ○ the relationship between the calling customer and the called customer, between the service-use date and the customer's birthday, etc.;
  ○ the service usage – the more you use, the cheaper the unit price.

SLA management:

- constantly monitors the SLA satisfaction level for each customer;
- generates warning information when the threshold is approached (input for setting up the service delivery priority level);
- generates penalty information for billing when the SLA is broken.

Service management needs to be developed for services/applications/content/ information registration, maintenance and removal, and for setting up the QoS requirements and security requirements.

Quality of service management needs to be developed to monitor the service quality delivered to the customer, to generate QoS-based billing information, and to predict the service quality trends.

Terminal equipment management needs to be developed for providing the CTE's capability, for reconfiguring the CTE and upgrading the software for the requested service, and for disabling stolen terminals remotely.

Security management needs to be further developed for AAA, for network access, for service access, for data confidentiality and for data integrity.

7.7.1 GRID Technologies

A GRID is a distributed resource. For NGN, GRID technology is about:

- resource sharing, virtualizing and orchestrating, where the resource can be network bandwidth, service capabilities, computing and storage;
- using large amounts of distributed network resources to form extremely high speed and high quality connections, e.g. a 6.4 Gbps TCP link across continent for astronomy or particle physics research.
- convergence of Web services and GRID.

In the NGN environment, the GRID technology can be applied in:

- the dynamic network resource management (for optimized resource utilization) for co-existing multiple access networks and multiple core transport technologies;
- mobile or wireless thin clients
- computing and storage for the distributed information system, e.g. HSS;
- internal usage, e.g. planning (organizing resources located in different places), billing, fault detection and traffic analysis;
- service hosting;
- content distribution.

7.7.2 End-to-End QoS Management

A possible end-to-end QoS management system is shown in Figure 7.14, which includes monitoring, analysing, reporting, policy-consulting and controlling, where:

- *Monitoring* monitors the availability and performance along the service delivery chain – this is different for user-to-user and for user-to-machine services,

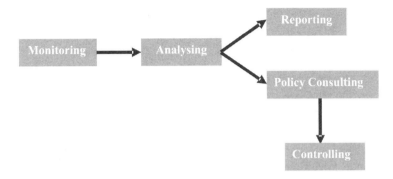

Figure 7.14   An end-to-end management system

- for user-to-user service, the service delivery chain is 'User → terminal → access network → core network → service platform → core network → access network → terminal → User';
- for user-to-machine, the service delivery chain is 'User → terminal → access network → core network → service platform → Server'.

The monitoring is done via the measurements:

- to measure as an end user,

  (a) if the targeted service functions or not,
  (b) if the pre-defined KQIs (key quality indicators) are reached or not.

- to measure the accumulated satisfaction level of SLs agreed in the SLA for each customer or aggregated customer;
- to measure the performance of servers, platforms, network nodes and connections against the pre-defined KPIs.

- *Analysing* analyses the collected data from monitoring to recognize a problem and to allocate the problem source, or to predict a potential problem or the performance tendency. Considering the complicated and twisted relationship between the KQIs delivered to the customer and the KPIs from the network and service platform, artificial intelligence provides the means for the analysing, including:

  - correlating the KQIs and KPIs in both a deterministic and a statistic way;
  - analysing the seriousness of each detected problem (monitored as end user), and allocating the possible problem sources in the network or service platform;
  - analysing the seriousness of each detected performance problem at radio, cable, network node, service node, etc., and estimating the impact on the services/applications, service level and end-user.

- *Reporting* generates the information for reporting to the relevant persons, e.g.

  ○ report to the management team about the current operation status, e.g. normal or problematic and when problematic,

  (a) the affected customer's number, geo-area,
  (b) the affected service/application/content/ information and extent,
  (c) the financial consequences, i.e. reduced income or penalty;

  ○ report to operating personnel about the detected problem and the urgency of repair (according to predefined criteria);
  ○ report to customer care about the problem that has occurred, a comprehensive explanation and the estimated repair time;
  ○ report to the billing system about satisfaction-based charging information for lower or no payment and SLA satisfaction-based charging information for penalty payment.

- *Policy consulting* consults with the policy server on what to do. 'Policy' is based on the operation strategy for allocating network resources according to business decisions. It instructs the actions of control mechanisms, is composed of predefined rules in the form of 'if condition = true, then action'. 'Policy' is required for end-to-end consistent control.
- *Controlling* is action according to the result of the policy consulting, which could be reactive or proactive. Compared with reactive control, proactive control requires prediction. Reactive controlling is to take action after a problem had appeared, following the steps of:

$$\text{Monitoring} \rightarrow \text{Analysis} \rightarrow \text{Policy consulting} \rightarrow \text{Action}$$

Proactive controlling is to take action before a problem has developed, following the steps:

$$\text{Monitoring} \rightarrow \text{Analysis} \rightarrow \text{Prediction} \rightarrow \text{Policy consulting} \rightarrow \text{Action.}$$

where the action is based on the current state predicting the danger of breaking the SL. The action can be:

○ to automatically repair the detected problem, e.g. to restart a server or to restart an application;
○ to change the RRM rules, e.g. to lower the SL for lower priority services/applications, to disconnect the lower priority user or to discard a lower priority packet;
○ to warn the operator for the need of new measures for maintaining the SLA.

7.7.3   End-to-End Security Management

As important as end-to-end QoS management, end-to-end security management should cover the following aspects:

- *authentication*, which is required to verify claimed identity;
- *authorization*, which enables certain actions after authentication and ensures that only authorized persons or devices are allowed access to network elements, services and applications;
- *confidentiality*, which is for data that should not be handled by unauthorized entities;
- *integrity*, ensuring that data is not modified during transition;
- *non-repudiation*, which ensures that the sender does not deny the origin of the received data;
- *communication security*, which ensures that information flows only between the authorized end points;
- *availability*, which ensures that there is no denial of authorized access to network elements, services or applications;
- *privacy security*, which provides data protection from disclosure to an authorized entity.

7.8   ADVANCED TECHNOLOGIES FOR NETWORK AND SERVICE MANAGEMENT

Considering the situation today, promising technologies that are still under development are Intelligent Agent, Artificial Intelligence SON and GRID.

7.8.1   Intelligent Agent Technology [14]

Intelligent Agent Technology has the potential to be used in many places for the network management including:

- producing an on-line active inventory for the network by sending the software agent to retrieve the latest information about the network topology, network node hardware and software;
- adjusting the transmitting power of a radio station by sending the software agent to negotiate when the generated interference from a neighbouring base station is too high;
- remote re-configuration of the network node with mobile agent according to the actual requirement for capacity distribution.

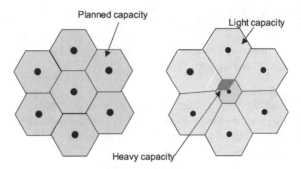

Figure 7.15 (Left) the original planned capacity distribution; (right) the mobile agent adapted capacity distribution by re-configuring node B

For example, in the WCDMA radio network shown in Figure 7.15, thanks to the mobile agent, which enables the re-configuration of a mobile base station, Node B, the network capacity is re-distributed according to the traffic demand. The left one indicates the original plan with homogeneous capacity distribution; the right one indicates the heterogeneous capacity distribution after the re-configuration done by mobile agent.

When approaching congestion, node B will send an intelligent agent to the neighbouring base station node Bs to negotiate the parameter configuration, e.g. pilot channel power, total transmission power, admission control criteria and handover criteria, in order to adjust the capacity distribution of the radio network.

7.8.2   Artificial Intelligence Technology [20–22]

Artificial intelligence technology is designed to be used for statistic data analysis to discover fuzzy co-relationships among multi-dimensional and enormous quantities of data, e.g. financial analyses or natural disaster predictions – both have thousands of parameters that influence the outcome but in a statistical or non-deterministic way.

The technology has also been used in the telecommunications industry to predict the customer churn rate by statistically analysing customer data covering customer gender, age, profession, income, residential location, native language, subscription history, terminal and service usage. Recently, the technology has also been used to enable self-configuration and self-optimization (or self-tuning) for a WCDMA radio network.

NGN promises an optimized usage of network resources and a guaranteed user experience. Artificial intelligence technology is a promising technology to play the role in both places for self-configuration and self-optimization of a radio network in a multiradio environment, and for proactive network performance management. Below are two examples to elaborate the approach.

Artificial Intelligence for Self-configuration and Self-optimization of a WCDMA Network

In a WCDMA network, there are thousands of parameters that need to be configured for each base station. Those parameters are dependent on the topology, morphology, radio environment, traffic demand pattern, network load and service type. Those parameters need to be correctly configured in order for the WCDMA network to perform in an optimized way. The complexity of a WCDMA network also comes from the fact that coverage, capacity and quality are coupled in a complex way, and there is no determined relationship between the configuration parameters and those network performance indicators.

The most difficult aspect is the dependency between service usage and network performance, which requires the constant optimization of a WCDMA network via configuration. Further more with the future home/office base station concept, the number of base stations in a radio network will be 10-fold or even 100-fold higher, and it will not be possible to configure such radio network in a traditional way.

The artificial intelligence technology will provide a statistical method to enable self-configuration by activating the loop as described below:

Step 1: cluster the worst performing base stations in terms of a single KPI or multiple KPIs.
Step 2: find the similarities in those worst performing base stations in terms of configuration parameters or topology, morphology, radio environment, traffic demand pattern, network load and service type.
Step 3: find the configuration parameters of the best performing base stations with similar topology, morphology, radio environment, traffic demand pattern, network load and service type.
Step 4: apply the configuration parameters of the best performing base stations to those worst performing base stations.

Going back to Step 1.

Artificial Intelligence for Proactive Network Performance Management

Today, the network performance management is passive, i.e. it acts after the performance problem has occurred. NGN promises guaranteed service accessibility at any time and at any place, which requires proactive performance management, i.e. to predict the possible performance problem and to activate the measure(s) before the performance problem occurs (Figure 7.16).

A tool based on artificial intelligence can function as follows:

• Build up primary knowledge about the statistic relationship between performance and condition (artificial intelligence technology is capable of learning

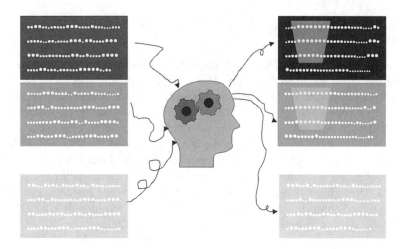

Figure 7.16 Artificial intelligence for data analysis of end-to-end network performance

from the histogram data the statistical relationship between the cause and consequence).

- Constantly monitor the network performance and condition to enrich knowledge about the statistical relationship between performance and condition.
- Constantly analyse the performance tendency according to the accumulated knowledge to predict the possible performance problems.
- Activate the measure(s) to avoid the performance problem happening.

7.8.3   SON Technology

SON (self-organizing networks) technology will automate network management and enable the reliability, scalability and flexibility of the future network. A SON-enabled network is capable of self-forming, self-planning, self- configuration, self- optimization and self-healing, where:

- self-forming is the automatic recognition of a network when a new node is added or an old node is removed;
- self-organizing is the node accepting or refusing the connection to any other nodes, routing traffic around congested mesh parts;
- self-planning is a radio base station or an access point's planning its coverage and capacity;
- self-configuration is a node configuring its system parameters, IP addresses and routing table;

- self-optimization is a node modifying its system parameters against the performance;
- self-healing is a network recovering from traffic congestion, a node problem or a link problem.

The SON related work is still in its infancy, and the realization of SON will be stepwise.

REFERENCES

[1] http://www.ngmn.org
[2] http://www.3gpp.org
[3] http://www.3gpp2.org
[4] http://www.itu.org
[5] http://www.ieee.org
[6] http://cordis.europa.en/FP6
[7] Eurescom Summit 2005 NFC Tutorial from Josef Noll and Juan Calvet.
[8] http://www.tmforum.org
[9] http://www.IETF.org
[10] http://www.biometricgroup.com
[11] http://www.nfc-forum.org
[12] http://www.Parlay.org
[13] http://www.OpenMobileAlliance.org
[14] http://www.ist-shuffle.org
[15] http://www.wireless-world-research.org
[16] https://www.ist-mobilife.org
[17] http://www.ambient-networks.org
[18] http://www.e2r.motlabs.com
[19] https://www.ist-winner.org
[20] Jaana Laiho *et al.*, *Radio Network Planning and Optimisation for UMTS*. Wiley: Chichester, 2002.
[21] Kevin Gurney, *An Introduction to Neural Networks*. CRC Press: Boca Raton, FL, 1997.
[22] Christopher M. Bishop, *Neural Networks for Pattern Recognition*. Oxford University Press: Oxford, 1995.

8

NGN Standardizations

The NGN standardization work started in 2003 within ITU-T, and is worldwide today in various major telecom standardization bodies. The most active NGN-relevant standardization bodies are ITU, ETSI, ATIS, CJK and TMF. The Next Generation Mobile Networks (NGMN) initiative is a major body for mobile-specific NGN activities, which are also becoming important contributors to the 3GPP specification for NGMN.

By end of 2005, ITU and ETSI had published the NGN specification Release 1:

- ITU NGN R1 defined the framework of NGN and was conceptual and not implementable.
- ETSI NGN R1 was targeted at implementable specifications, although focusing on VoIP and xDSL access.

Both bodies are continuing their NGN work for the further releases.

ATIS and CJK have made important contributions to the ITU NGN specification. TMF has been working on NGN management concepts and tools under the name NGOSS (New Generation Operation System and Software). Part of NGOSS has been adopted by both ITU and ETSI.

The NGMN Alliance (former NGMN Initiative), founded in 2006, is for the commercial launch of a new experience in mobile broadband communication beyond 2010. It has been extended and become extremely active since the 3GSM world congress 2007 in February 2007. The NGMN Alliance is not a standardization body; however, it has such an impact on 3GPP specification development for LTE (long-term evolution) and SAE (system architecture

Next Generation Networks: Perspectives and Potentials Jingming Li Salina and Pascal Salina
© 2007 John Wiley & Sons, Ltd

evolution). 3GPP LTE and SAE will form the so-called 3.9G mobile system with 4G performance using 3G spectrum that guarantees smooth evolution to 4G.

8.1   ITU AND GSI-NGN

The ITU (International Telecommunication Union, http://www.itu.org) is an international organization within the United Nations in which governments and the private sector coordinate global telecom networks and services. ITU-T is the telecommunications sector of ITU. Its mission is to produce high-quality recommendations covering all the fields of telecommunications.

In 2003, under the name JRG-NGN (Joint Rapporteur Group on NGN), the NGN pioneer work was initiated. The key study topics are [1,2]:

- NGN requirements;
- the general reference model;
- functional requirements and architecture of the NGN;
- evolution to NGN.

With another 11 draft texts, two fundamental recommendations on NGN have been produced and approved, which are:

- Y.2001: 'General overview of NGN'.
- Y.2011: 'General principles and general reference model for next-generation networks'.

These two excellent documents comprise the basic concept and definition of NGN.

In May 2004, the FG-NGN (Focus Group on Next Generation Networks) was established in order to continue and accelerate NGN activities initiated by the JRG-NGN. FG-NGN addressed the urgent need for an initial suite of global standards for NGN. The NGN standardization work was launched and mandated to FG-NGN.

On 18 November 2005, the ITU-T published its NGN specification Release 1, which is the first global standard of NGN and marked a milestone in ITU's work on NGN. The NGN specification Release 1, with 30 documents, specified the NGN Framework, including the key features, functional architecture, component view, network evolution, etc. Lacking protocol specifications, the ITU NGN Release 1 is not at an implementable stage; however, it is clear enough to guide the evolution of today's telecom networks. With the release of NGN Release 1, the FG-NGN has fulfilled its mission and closed.

Following FG-NGN, the ITU-T NGN standardization work continues under the name GSI-NGN (NGN Global Standards Initiative) in order to maintain and develop the FG-NGN momentum. Today, the GSI-NGN work is on-going at full speed. In parallel with the FG-NGN, there are two other groups working on the NGN relevant issues. They are the NGN-MFG (NGN Management Focus Group) and the OCAF-FG (Open Communication Architecture Forum Focus Group).

Since 2004, NGN-MFG has been working on the specification of NGN management. The NGN-MFG is targeted at:

- developing a set of interoperable specifications as solution for the management of NGN services and networks;
- developing an NGN management specification roadmap for NGN Release 1.

Since 2004, OCAF-FG has been working on the specification of NGN for open communication architecture. The OCAF-FG is targeted at agreed-on specifications for a set of components for a new carrier-grade open platform that will accelerate the deployment of NGN infrastructure and services. Today, both NGN-MFG and OCAF-FG contribute directly to the GSI-NGN.

8.1.1 GSI-NGN Concept [Reproduced with the Permission of ITU]

NGN Definition

The ITU has defined the NGN as 'A packet-based network able to provide telecommunications services and able to make use of multi broadband, QoS-enabled transport technologies and in which service related functions are independent from underlying transport-related technologies. It offers unfettered access by users to different service providers. It supports generalized mobility which will allow consistent and ubiquitous provision of services to users'.

NGN Key Features

The ITU's NGN possesses the following key features:

- packet-based transfer;
- separation of control functions among bearer capabilities, call/sessions and applications/services;
- decoupling of service provision from transport, and provision of open interfaces;
- support for a wide range of services, applications and mechanisms based on service building blocks (including real time/streaming/noon-real time services and multimedia);

- broadband capabilities with end-to-end QoS;
- interworking with legacy networks via open interfaces;
- generalized mobility;
- unfettered access by users to different service providers;
- a variety of identification schemes;
- unified service characteristics for the same service as perceived by the user;
- converged services between fixed/mobile;
- independence of service-related functions from underlying transport technologies;
- support of multiple last mile technologies;
- compliance with all regulatory requirements, e.g. concerning emergency communications, security and privacy.

Functional Architecture [3]

The current functional architecture of ITU NGN is designed to support so-called Release 1 services and Release 1 requirements (Figure 8.1); a realization of this functional architecture with component view is shown in Figure 8.2. This

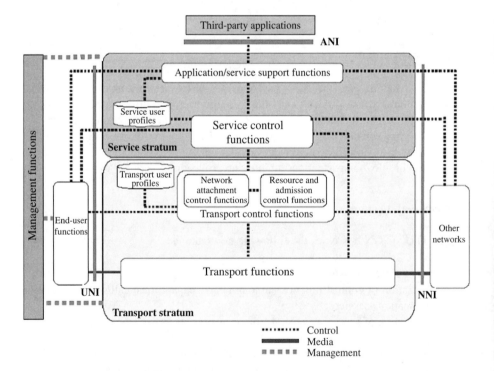

Figure 8.1 GSI-NGN functional architecture. Reproduced by kind permission of ITU

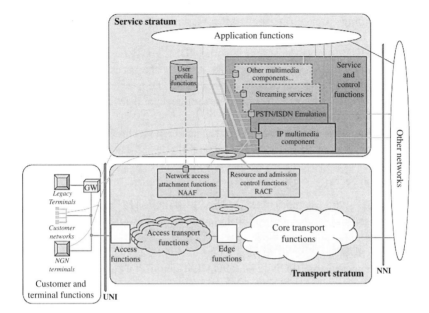

Figure 8.2 Component view of a possible realization of GSI-NGN functional architecture. Reproduced by kind permission of ITU

functional architecture is composed of functional groups separated by well-defined interfaces. Each functional group contains a set of functional entities.

The main functional groups are the transport stratum, the service stratum, the end-user functions, the third-party applications, the management functions and the other networks. The main interfaces are the UNI between the user and network interfaces, the ANI between the application and network interfaces and the NNI between the network and network interface.

The solid lines indicate the user traffic; the dashed lines indicate the signalling paths; the thick dashed lines indicate the management data flows.

Transport Stratum Functions

The transport stratum functions include transport functions, transport control functions and transport user profiles.

Transport functions provide the connectivity for all components and physically separated functions within the NGN. These functions provide support for the transfer of media information, as well as the transfer of control and management information. Transport functions include access transport functions, edge functions, core transport functions and gateway functions.

The *access transport functions* take care of end-users' access to the network as well as collecting and aggregating the traffic coming from these accesses towards the core network. These functions also perform QoS control mechanisms dealing directly with user traffic, including buffer management, queuing and scheduling, packet filtering, traffic classification, marking, policing and shaping. These functions also include access-technology dependent functions, e.g. the WCDMA mobile access and the xDSL fixed access. Depending on the technology used for accessing NGN services, the access network includes functions related to optical access, cable access, xDSL access, wireless access, e.g. IEEE 802.11 and 802.16 access technologies, and IMT2000 radio access technologies.

The following is a non-exhaustive list of candidate technologies to implement access transport functions for NGN Release 1.

- Wireline access:

  o xDSL – this includes ADSL [15,17], SHDSL [18] and VDSL [19,20] transport systems and supporting connection/multiplexing technologies;
  o SDH dedicated bandwidth access [21];
  o optical access – this covers point-to-point [22] and xPON transport systems such as BPON [23], GPON [24] and EPON [25] (Gigabit EPON is sometimes called GEPON).
  o cable networks – this covers cable networks based on packet cable multimedia specifications [26].
  o LANs – this covers LANs using either coaxial or twisted pair cable, including 10Base-T ethernet [27], fast ethernet [28], gigabit ethernet [29] and 10 gigabit ethernet [30];
  o PLC networks – the PLC network transmits and receives data over the power line.

- Wireless access:

  o IEEE 802.11x WLAN;
  o IEEE 802.16x WiMAX;
  o any 3GPP or 3GPP2 IP-CAN (NGN does not support the CS domain as an access transport technology);
  o broadcast networks – this covers 3GPP/3GPP2 Internet broadcast/ multicast, DVB, and ISDB-T.

The *edge functions* are used for media and traffic processing when aggregated traffic coming from different access networks is merged into the core transport network; they include functions related to support for QoS and traffic control. These functions are also used between core transport networks.

The *core transport functions* are responsible for ensuring information transport throughout the core network. These functions provide IP connectivity, at the transport stratum and across the core network, and provide the means to differentiate the quality of transport in the core network. They also provide QoS mechanisms dealing directly with user traffic, including buffer management, queuing and scheduling, packet filtering, traffic classification, marking, policing, shaping, gate control, and firewall capability.

The *gateway functions* provide capabilities to interwork with end-user functions and other networks, including other types of NGN and many existing networks, such as the PSTN/ISDN and the public Internet. These functions can be controlled either directly from the service control functions or through the transport control functions.

The *media handling functions* provide media resource processing for service provision, such as the generation of tone signals and trans-coding. These functions are specific to media resource handling in the transport stratum.

The *transport control functions* include resource and admission control functions (RACF) and network attachment control functions (NACF). The RACF provides QoS control (including resource reservation, admission control and gate control), NAPT and/or FW traversal control functions over access and core transport networks. The Admission control involves checking authorization based on user profiles, SLAs, operator-specific policy rules, service priority and resource availability within access and core transport.

The RACF acts as the arbitrator for resource negotiation and allocation between service control functions and transport functions. The RACF interacts with service control functions and transport functions for session-based applications (e.g. SIP call) and non-session-based applications (e.g. video streaming) that require the control of NGN transport resource, including QoS control and NAPT/FW control and NAT traversal. The RACF interacts with transport functions for the purpose of controlling one or more the following functions in the transport layer: packet filtering; traffic classification, marking, policing and priority handling; bandwidth reservation and allocation; network address and port translation; and firewall. The RACF interacts with NACF, including network access registration, authentication and authorization, and parameter configuration for checking user profiles and SLAs held by them. For those services across multiple providers or operators, service control functions, the RACF and transport functions may interact with the corresponding functions in other packet networks.

The NACF provides registration at the access level and initialization of end-user functions for accessing NGN services. These functions provide network-level identification/authentication, manage the IP address space of the access network and authenticate access sessions. These functions also announce the contact point of NGN service/application support functions to the end user. The NACF provides further the functionality of:

- dynamic provision of IP addresses and other user equipment configuration parameters;
- authentication at the IP layer (and possibly other layers);
- authorization of network access, based on user profiles;
- access network configuration, based on user profiles;
- location management at the IP layer.

The *transport user profile functions* take the form of a functional database representing the combination of a user's information and other control data into a single 'user profile' function in the transport stratum. This functional database may be specified and implemented as a set of cooperating databases with functionalities residing in any part of the NGN.

The *service stratum functions* include the service control functions and the application/service support functions, as well as service user profile functions.

The *service control functions* include both session and non-session control, registration and authentication and authorization functions at the service level. They can also include functions for controlling media resources, i.e. specialized resources and gateways at the service-signalling level.

Within the service control functions, the possible components included are the IP multimedia component, the PSTN/ISDN emulation component, the streaming services component and other multimedia components.

The IP multimedia service component is a service component based on the capabilities of the 3GPP IP Multimedia Subsystem (IMS) [3]. It has been a starting point for the definition of Release 1 to leverage the capabilities of the 3GPP IMS. The IMS functionality for NGN Release 1 employs SIP-based service control [4]. To support the heterogeneous access transport environment of Release 1 the capabilities of the 3GPP IMS need to be extended. NGN Release 1 will maintain full compatibility with 3GPP/3GPP2 IP connectivity access transport functions (e.g. IP-CAN) and terminals.

The *PSTN/ISDN emulation service component* is a service component defined to support PSTN/ISDN replacement scenarios, with full interoperability with existing (legacy) PSTN/ISDN networks. This component fully supports legacy (PSTN/ISDN) interfaces to customer equipment and provides the user with identical services and experience to that of the existing PSTN/ISDN.

The *application/service support functions* include functions such as the gateway, registration, authentication and authorization functions at the application level. These functions are available to the 'third-party applications' and 'end-user' functional groups. The application/service support functions work in conjunction with the service control functions to provide end-users and third-party application providers with the value-added services they request.

Through the UNI, the application/service support functions provide a reference point to the end-user functions, e.g. in the case of third-party call control for click to call service. The third-party applications' interactions with the application/ service support functions are handled through the ANI reference point.

NGN will help in the creation and offering of new services. As the number, sophistication and degree of interworking between services increase, there will be a need to provide more efficiency and scalability for network services. Therefore, NGN applications and user services should be able to use a flexible service and application-provisioning framework.

Such a framework should enable application providers, both NGN internal and third-party, to implement value-added services that make use of network capabilities in an agnostic fashion. Network capabilities and resources that are offered to applications are defined in terms of a set of capabilities inside this framework and are offered to third-party applications through the use of a standard application network interface. This provides a consistent method of gaining access to network capabilities and resources, and application developers can rely on this consistency when designing new applications. The internal NGN application providers can make use of the same network capabilities and resources that are used by third-party application providers.

NGN Release 1 should support the following three classes of value-added service environments:

• IN-based service environment – support for intelligent network (IN) services. Examples of ANI-specific protocols for this environment include IN Application Protocol [5], Customised Application for Mobile network Enhanced Logic (CAMEL) [6,7] and Wireless Intelligent Network (WIN) [8].
• IMS-based service environment – support for IMS-based service environment. Examples of ANI-specific interfaces include ISC, Sh, Dh, Ut, Ro, Rf, Gm and Mb [9].
• Open service environment – support for open service environments. Examples of this environment using ANI include OSA/Parlay, Parlay X and OMA [10,13].

The *service user profile functions* represent the combination of user information and other control data into a single user profile function in the service stratum, in the form of a functional database. This functional database may be specified and implemented as a set of cooperating databases with functionalities residing in any part of the NGN.

Release 1 defines the user profile functions, which provide capabilities for managing user profiles and making the user profile information available to other NGN functions. A user profile is a set of attribute information related to a user. The user profile functions provide the flexibility to handle a wide variety of user information. Some of the user profile models that may inform the design of the user profile functions include:

• 3GPP Generic User Profile (GUP);
• 3GPP2 User Profile;
• W3C Composite Capabilities/Preference Profile (CC/PP);
• OMA User Agent Profile;
• 3GPP/ETSI Virtual Home Environment;
• Parlay Group – user profile data.

As shown in Figure 8.2, the user profile functions support the identified service and control functions in the service stratum, as well as the network access attachment functions in the transport stratum. This central role for the user profile functions is natural, since users and their service requirements are the driving forces behind the existence of the network itself.

End-user Functions

No assumptions are made about the diverse end-user interfaces and end-user networks that may be connected to the NGN access network. Different categories of end-user equipment are supported in the NGN, from single-line legacy telephones to complex corporate networks. End-user equipment may be either mobile or fixed.

Customers may deploy a variety of network configurations, both wired and wireless, inside their customer network. This implies, for example, that Release 1 will support simultaneous access to NGN through a single network termination from multiple terminals connected via a customer network.

It is recognized that many customers deploy firewalls and private IP addresses in combination with NAPT. NGN support for user functions is limited to control of (part of) the gateway functions between the end user functions and the access transport functions. The device implementing these gateway functions may be customer or access transport provider-managed. Management of customer networks is, however, outside the scope of Release 1. As a result, customer networks may have a negative impact on the QoS of an NGN service as delivered to user equipment.

Implications of specific architectures of customer networks on the NGN are beyond the scope of Release 1. Customer network internal communications do not necessarily require the involvement of the NGN transport functions (e.g. IP PBX for corporate network).

User Equipment

The NGN should support a variety of user equipment. This includes gateway and legacy terminals (e.g. voice telephones, facsimile, PSTN textphones etc.), SIP phones, soft-phones (PC programmes), IP phones with text capabilities, set-top boxes, multimedia terminals, PCs, user equipment with an intrinsic capability to support a simple service set and user equipment that can support a programmable service set.

It is not intended to specify or mandate a particular NGN user equipment type or capability, beyond compatibility with NGN authentication, control and transport protocol stacks.

NGN supports a mobile terminal that is fully compliant with 3GPP specifications only when directly connected through a 3GPP IP-CAN. Release 1 may not support 3GPP mobile terminals when they are not directly connected through a 3GPP IP-CAN.

Release 1 should allow the simultaneous use of multiple types of access transport functions by a single terminal; however there is no requirement to coordinate the communication. Such terminals may therefore appear to be two or more distinct terminals from the network point of view.

The user equipment should enable interface adaptation to varying user requirements, including the needs of people with disabilities, for connection with commonly provided user interface devices.

Management Functions

Support for management is fundamental to the operation of the NGN. These functions provide the ability to manage the NGN in order to provide NGN services with the expected quality, security and reliability. These functions are allocated in a distributed manner to each functional entity (FE), and they interact with network element (NE) management, network management and service management FEs. Further details of the management functions, including their division into administrative domains, can be found in ITU-T M.3060 [2].

The management functions apply to the NGN service and transport strata. For each of these strata, they cover the following areas:

- fault management;
- configuration management;
- accounting management;
- performance management;
- security management.

The accounting management functions also include charging and billing functions. These interact with each other in the NGN to collect accounting information, in order to provide the NGN service provider with appropriate resource utilization data, enabling the service provider to properly bill the users of the system.

Network Node Interfaces

Interconnection and network node interfaces (NNIs) – as well as interconnection between multiple NGN administrative domains, the NGN is also required to support access to and from other networks that provide communications, services and content, including the secure and safe interconnection to the Internet.

NGN Release 1 provides support for services across multiple NGN administrative domains. Interoperability between NGN administrative domains shall be based on defined interconnect specifications.

Table 8.1 Release 1 (P-)NNIs for interconnection to other types of networks. Reproduced by kind permission of ITU

Type of networks	Signalling interface	Bearer interface
Circuit-based networks	ISUP	TDM
IP-based networks	SIP (session control)	IPv4
	IPv4	IPv6
	IPv6	MIPv4
	MIPv4	MIPv6
	MIPv6	RTP
	BGP	RTCP
	HTTP	

NNIs to non-NGNs – Release 1 supports interconnection to other IP networks and by implication to any IP-based network that complies with the NGN interconnection protocol suite. It supports direct interconnection with the PSTN/ISDN by means of interworking functions that are implemented within the NGN. Interoperability between NGN and non-NGN will be based on defined interconnect specifications.

Table 8.1 lists the candidate interconnection interfaces, including a non-exhaustive list of protocols that may be supported in Release 1 and may be applied as P-NNIs to Enterprise networks. The following is the list of candidate networks that will interconnect using NNIs to the NGN:

- Internet;
- cable networks;
- enterprise networks;
- broadcast networks;
- PLMN networks;
- PSTN/ISDN networks.

NNIs between NGNs – NGN release 1 allows for the partition of the NGN into separate administrative domains. Interfaces on a trust boundary between domains need to support various functionalities to enable robust, secure, scaleable, billable, QoS-enabled and service-transparent interconnection arrangements between network providers. Some of the trusted domain's internal information may be removed across a trust boundary, for instance to hide the user's private identity or network topology information.

8.1.2 GSI-NGN Release 1

The basic achievements of GSI-NGN Release 1 can be summarized as follows:

- NGN principles, Release 1 Scope;
- high-level requirements and capabilities (stage 1);

- high-level architecture, some components in detail (stage 2);
- some capabilities in detail (stages 1 and 2) – QoS, security, mobility.

8.1.3 GSI-NGN Release 2

GSI-NGN Release 2 is planned to achieve:

- high-level requirements and capabilities – start (stage 1);
- high-level/component architecture evolution – start (stage 2);
- service-specific scenarios, requirements and capabilities (stage 1).

8.1.4 NGN Recommendations

As of 15 November 2006, the NGN recommendations are listed in Table 8.2.

8.2 ETSI AND TISPAN-NGN

ETSI (European Telecommunications Standards Institute, http://www.etsi.org) is a standard organization active in all areas of telecommunications (radio communications, broadcasting and information technologies). Its mission is to produce telecommunications standards for today and for the future. ETSI is also on the way to becoming the home of ICT standardization for Europe. ETSI also contributes to the ITU standardization.

In May 2003, ETSI formed the TISPAN (Telecommunications and Internet-converged Services and Protocols for Advanced Networking) project targeted at specifying NGN. In December 2005, ETSI published TISPAN NGN specification Release 1 and the further release plan. According to ETSI:

- The TISPAN NGN specification Release 1, provides the first set of implementable NGN specifications, although it focuses on VoIP and xDSL access. It is now being used by industry to build the NGN.
- The TISPAN NGN specification Release 2 is on-going with the focus on enhanced mobility, new services and content delivery with improved security and network management. It is planned to be completed in 2007, and Release 3 in 2009.

Today, TISPAN and 3GPP are working together to define a harmonized IMS-centric core for wireless and wireline networks.

The timescales and dependencies of ETSI NGN Release 1 are aligned with 3GPP Release 7 work on FBI, and ETSI NGN Release 2 with 3GPP Release 8 work on FMC.

Table 8.2 ITU-T GSI NGN recommendation list

Group name	Number	Title
Y.2000–Y.2099 Frameworks and functional architecture models	Y.2091 (Y.term)	Terms and definitions for NGN
	Y.2201 (Y.NGN-R1-Reqts)	NGN Release 1 requirements
	Y.2001	General overview of NGN
	Y.2011	General principles and general reference model for next-generation networks
	Y.2012 (Y.NGN-FRA)	Functional requirements and architecture of the NGN
	Y.2021 (Y.IFN)	IMS for next-generation networks
	Y.2031 (Y.PIEA)	PSTN/ISDN emulation architecture
	Y.2091 (Y.term)	Terms and definitions for next-generation networks
Y.2100–Y.2199 Quality of service and performance	Y.2111 (Y.RACF)	Resource and admission control functions in next-generation networks
	Y.2171 (Y.CACPriority)	Admission control priority levels in next-generation networks
Y.2200–Y.2249 Service aspects: service capability and service architecture	Y.2201(Y.NGN-R1-Reqts)	NGN release 1 requirements
Y.2250–Y.2299 Service aspects: interoperability of services and networks in NGN	Y.2261 (Y.piev)	PSTN/ISDN evolution to NGN
	Y.2262 (Y.emsim)	
	Y.2271 (Y.csem)	Cak server based PSTN/ISDN emulation
Y.2300–Y.2399 Numbering, naming and addressing		

Y.2400–Y.2499 Network management	Y. 2401	Principles for the management of the next-generation networks
Y.2500–Y.2599 Network control architectures and protocols		
Y.2600–Y.2699 Reserved	Y.2601 (Y.FPBN-req) Y.2611 (Y.FPBN-arch)	
Y.2700–Y.2799 Security	Y.2701 (Y. NGN Security)	Security requirements for NGN Release 1
Y.2800–Y. 2899 Generalized mobility	Y.2801 = Q.1706	
Y.2900–Y.2999 Reserved		

8.2.1 TISPAN-NGN Concept [Reproduced with the Permission of ETSI] [34]

The TISPAN-NGN is targeted at:

- Providing NGN services

  o conversation (voice call, video call, chat, multimedia sessions);
  o messaging (email, SMS, EMS, MMS, instant messaging and presence);
  o content-on-demand (browsing, download, streaming, push, broadcast).

- Supporting access technologies

  o 3GPP standardized mobile GSM/GPRS/EDGE/UMTS/HSPA/LTE;
  o fixed DSL;
  o wired LAN;
  o wireless LAN;
  o cable

The TISPAN NGN specification covers NGN services, architectures, protocols, QoS, security and mobility aspects within fixed networks. TISPAN and 3GPP are working together to define a harmonized IMS-centred core for both wireless and wireline networks. This harmonized all-IP network has the potential to provide a completely new telecom business model for both fixed and mobile network operators. Access-independent IMS will be a key enabler for fixed/mobile convergence, reducing network installation and maintenance costs, and allowing new services to be rapidly developed and deployed to satisfy new market demands.

Figures 8.3–8.5 provides an overview of the NGN architecture. This NGN functional architecture described complies with the ITU-T general reference model for next-generation networks [3] and is structured according to a service layer and an IP-based transport layer.

Service Layer

The service layer comprises the following components:

- core IP multimedia subsystem (IMS) – this component supports the provision of SIP-based multimedia services to NGN terminals and also supports the provision of PSTN/ISDN simulation services;
- PSTN/ISDN emulation subsystem (PES) – this component supports the emulation of PSTN/ISDN services for legacy terminals connected to the NGN, through residential gateways or access gateways;
- streaming subsystem – this component supports the provision of RTSP-based streaming services to NGN terminals;

- content broadcasting subsystem – this component supports the broadcasting of multimedia content (e.g. movies, television channels etc.) to groups of NGN terminals;
- common components – the NGN architecture includes a number of functional entities that can be accessed by more than one subsystem. As shown in Figure 8.6, these are:

  o the user profile server functions (UPSF);
  o the subscription locator function (SLF);
  o the application server function (ASF);
  o the interworking function (IWF);
  o the interconnection border control function (IBCF);
  o the charging and data collection functions.

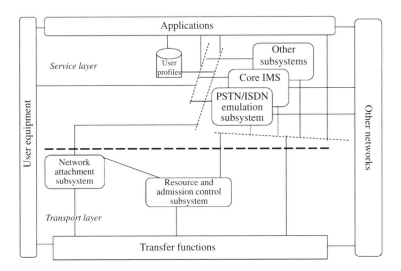

Figure 8.3   TISPAN-NGN overall architecture

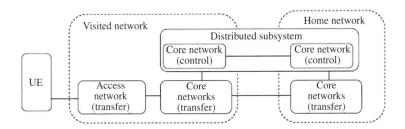

AQ1

Figure 8.4   Distributed subsystem between a visited and a home network

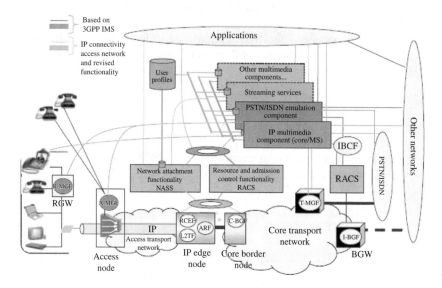

Figure 8.5   ETSI TISPAN-NGN example architecture with xDSL access

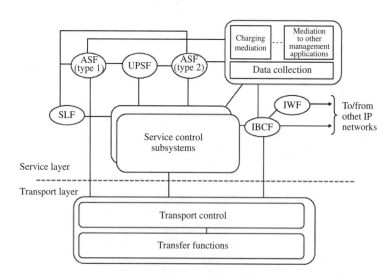

Figure 8.6   Common components overview

Transport Layer

The transport layer comprises a transport control sub-layer on top of transfer functions. The *transport control sub-layer* is further divided in two subsystems, i.e. the network attachment subsystem (NASS) and the resource and admission control subsystem (RACS).

NASS provides the following functionalities:

- dynamic provision of IP addresses and other terminal configuration parameters;
- authentication taking place at the IP layer, prior to or during the address allocation procedure;
- authorization of network access based on user profiles;
- access network configuration based on user profiles;
- location management taking place at the IP layer.

RACS provides admission control and gate control functionalities including the control of NAPT and priority making. Admission control involves checking authorization based on user profiles held in the access network attachment subsystem, on operator-specific policy rules and on resource availability. Checking resource availability implies that the admission control function verifies whether the requested bandwidth is compatible with both the subscribed bandwidth and the amount of bandwidth already used by the same user on the same access, and possibly other users sharing the same resources.

Figure 8.7 provides an overview of the *transfer functions* and their relationship with the other components of the architecture. Modelling of transfer functions here is limited to aspects that are visible to other components of the architecture. Only the functional entities that may interact with the transport control sub-layer or the service layer are visible in the transfer sub layer. These are:

- The media gateway function (MGF).
- The MGF provides the media mapping and/or transcoding functions between an IP-transport domain and switched circuit network facilities (trunks, loops). It may also perform media conferencing and send tones and announcements;.
- The border gateway function (BGF) provides the interface between two IP-transport domains. It may reside at the boundary between an access network and the customer terminal equipment, between an access network and a core network or between two core networks.
- The access relay function (ARF) acts as a relay between the user equipment and the NASS. It receives network access requests from the user equipment and forwards them to the NASS. Before forwarding a request, the ARF may

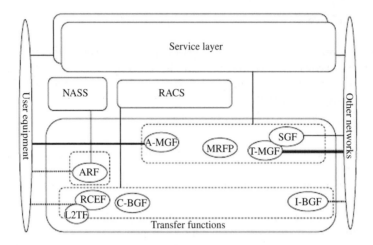

Figure 8.7 Transfer functions overview

also insert local configuration information and apply protocol conversion procedures.

- The signalling gateway function (SGF) performs signalling conversion (both ways) at the transport level between the SS7-based transport of signalling and IP-based signalling transport.
- The media resource function processor (MRFP) provides specialized resource processing functions beyond those available in media gateway functions. This includes resources for supporting multimedia conferences, sourcing multimedia announcements, implementing IVR (interactive voice response) capabilities and media content analysis.
- layer 2 termination function (L2TF).

An example of realization of this functional architecture, with an xDSL-based access network, is given in Figure 8.5. The configuration assumes the following:

- A border gateway function (C-BGF) is implemented in a core border node sitting at the boundary between the access network and a core network, at the core network side.
- A resource control and enforcement function (RCEF) is implemented in an IP edge node sitting at the boundary between core networks, at the access network side. In this example, this node also implements the L2TF and ARF functional entities.
- A border gateway function (I-BGF) is implemented in a border gateway (BGW) sitting at the boundary with other IP networks.

- A media gateway function (T-MGF) is implemented in a trunking media gateway (TGW) at the boundary between the core network and the PSTN/ISDN.
- A media gateway function (A-MGF) is implemented in an access node (AN), which also implements a DSLAM.
- A media gateway function (R-MGF) is implemented in a residential media gateway (RGW) located in the customer premises.

User Equipment

The user equipment (UE) consists of one or more user-controlled devices allowing a user to access services delivered by NGN networks. Different components of the customer equipment may be involved depending on the subsystem they interact with.

The UE functionalities are:

- Authentication – as shown in Figure 8.8, two levels of network iden-tification/authentication are available in the NGN architecture, namely at the level of the network attachment between UE and NASS and at the service layer level between NGN service control subsystems and applications.
- Interfaces:
  - Interfaces to the core IMS – access to the services of the IMS is provided to SIP-based terminals;.
  - Interfaces to the PSTN/ISDN emulation subsystem – access to the services of the PSTN/ISDN emulation subsystem is provided by legacy terminals through a gateway function, which may reside in customer premises or in the operator's domain.
  - Interfaces with applications – interactions with SIP application servers take place through the Ut interface. This interface enables the user to

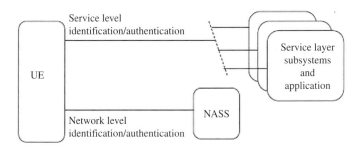

Figure 8.8 NGN authentication levels

manage information related to his or her services, such as creation and assignment of public service identities, management of authorization policies that are used, for example, by presence services or conference policy management.

o Interfaces with the NASS – these interfaces enable the user equipment to attach to the network and receive configuration information. Signalling between the UE and the NASS may be relayed via the ARF in the transfer sub-layer.

o Interface with RACS.

• Interconnection with other networks/domains – the interconnection can happen at the transport layer or at the service layer:

o interconnection at the transport layer;

(a) at the transfer layer – interconnection at the transfer level takes place either with TDM-based networks through T-MGF and SGF entities or with IP-based networks, at the Iz reference point, through an I-BGF entity (see Figure 8.9). Interconnection with SS7-based networks only applies to the IMS and PSTN/ISDN emulation subsystems. In such cases, the service layer controls the T-MGF entity behaviour. Interconnection with IP-based networks depends on the subsystems involved. The I-BGF may behave autonomously or under the control of the service layer, through the RACS, for services that involve the IMS core component or the PSTN/ISDN emulation subsystem. Future releases of the TISPAN specifications will address the control of the I-BGF in other configurations.

(b) at NASS;

(c) at RACS.

o Interconnection at the service layer – interconnection at the service layer can take place either with SS7-based networks or with IP-based networks. Interconnection with SS7-based networks only applies to the IMS and PSTN/ISDN emulation subsystems, both of which include appropriate functionality to interact with the T-MGF and the SGF. Interconnection with IP-based networks depends on the subsystems involved. IP-based interconnection to/from the IMS core component or the PSTN/ISDN emulation subsystem is performed using the IBCF entity and possibly the IWF entity (see Figure 8.10). Direct interconnection between other types of subsystems or applications is outside the scope of TISPAN R1. IP-based interconnection with external networks supporting a TISPAN-compatible version of SIP is performed at the Ic reference point, via the IBCF. Interconnection with external networks supporting H.323 or a non-compatible version of SIP is performed at the Iw reference point, via the IWF. The IBCF and the IWF communicate via the Ib reference point.

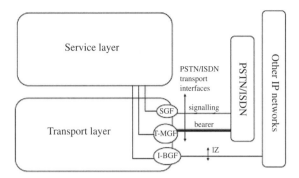

Figure 8.9 Network interconnection at transfer level

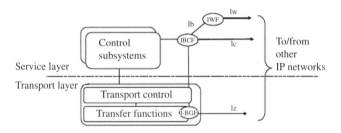

Figure 8.10 IP interconnection

8.2.2 TISPAN-NGN Release 1

The TISPAN-NGN Release 1 is focused on voice, SIP and xDLS access. Different from ITU-T GSI-NGN Release 1, ETSI TISPAN-NGN Release 1 is claimed to be implementable, providing robust and open standards which industry can use as a reliable basis for the development and implementation of the first-generation NGN system.

TISPAN-NGN Release 1 Scope

- Service capabilities:
  - real-time conversational services, e.g. voice call, video call, sessions;
  - messaging and presence, e.g. MMS, instant messaging;
  - PSTN/ISDN emulation and simulation.

- Network architecture:

  - 3GPP R6 IMS for SIP-based service control;
  - xDSL access;
  - initial network-edge QoS controls.

- IP connectivity subsystems:

  - NASS (network attachment subsystem) for xDSL access technology; IP address provisioning; network access authorization; user location management;
  - RACS (resource and admission control subsystem) for service-based local policy control; resource reservations; admission control; NAPT/Gate control.

- Service-oriented subsystems:

  - 3GPP R6 IMS supporting multimedia services and adapted to accommodate xDSL-based access, and PSTN/ISDN emulation and simulation;
  - PSTN/ISDN emulation subsystem specially tailored to allow TDM equipment replacement, while keeping legacy terminals unchanged.

- PSTN/ISDN emulation.
- Charging.
- Emergency communication.
- Service enablers.

TISPAN-NGN Release 1 Deliverables

The TISPAN-NGN Release 1 deliverables are listed in Table 8.3.

8.2.3   TISPAN-NGN Release 2

The on-going TISPAN-NGN R2 is focused on enhanced mobility, new services and content delivery with improved security. The main new working items in R2 are:

- requirements analysis for FMC;
- requirements analysis for home networking;
- requirements for network capabilities to support IPTV services;
- IPTV integration of NGN services and capabilities using IMS;
- support of business services and enterprise network interworking.

Table 8.3 TISPAN-NGN Release 1 deliverables

Group name	Number	Title
Overall NGN	TR 180 001	Release definition
	TR 180 000	Terminology
	TS 181 005	Services and capabilities requirements for TISPAN NGN Release 1
	ES 282 001	NGN architecture: overall network architecture; functional architecture
	TR 182 005	Organization of user data
Service related	TS 181 002	Multimedia telephony with PSTN/ISDN supplementary services
	TS 181 005	IP multimedia system (IMS) messaging; stage 1
	TS 181 001	Video telephony over NGN; service description
	TS 181 005	Presence service; stage 1
	EG 201 998-3	Service provider access; open service access for API requirements; part 3
	TS 182 011	XML group management; architecture and functional description (endorsement of OMA-AD-XDM-V1_0)
	TS 182 008	Presence service; functional architecture and functional description; 3GPP TS 23.141, modified; OMA-AD-Presence_SIMPLE-V1_0, modified (stage 2)
	TS 183 028	Common basic communication procedures
	TS 183 004	PSTN/ISDN simulation services; communication diversion (CDIV); protocol specification
	TS 183 005	PSTN/ISDN simulation services; conference (CONF); protocol specification
	TS 183 006	PSTN/ISDN simulation services; messaging waiting indications (MWI); protocol specification
	TS 183 007	PSTN/ISDN simulation services; Originating Identification Presentation (OIP) and Originating Identification Restriction (OIR); Protocol specification
	TS 183 008	PSTN/ISDN simulation services; terminating identification presentation (TIP) and terminating identification restriction (TIR); protocol specification
	TS 183 010	PSTN/ISDN simulation services; communication hold (HOLD); protocol specification
	TS 183 011	PSTN/ISDN simulation services; anonymous communication rejection (ACR) and communication barring (CB); protocol specification
	TS 183 016	PSTN/ISDN simulation services; malicious communication identification (MCID); protocol specification

Table 8.3 (Continued)

Group name	Number	Title
	ES 283 030	Presence service; protocol specification; 3GPP TS 24.141, modified; OMA-TS-XDM_Core-V1_0, modified (stage 3)
	TS 183 038	XML group management; stage 3 specification (endorsement of OMA OMA-TS-XDM_Core-V1_0-20050628 and OMA-TS-XDM-XDM_Shared-V1_0-20050628)
	TS 183 041	Messaging service using the IP multimedia (IM) core network (CN) subsystem; stage 3; protocol specifications
	TS 183 023	PSTN/ISDN simulation services; extensible markup language (XML) (XCAP) for manipulating NGN PSTN/ISDN simulation services; protocol specification
	TS 183 029	PSTN/ISDN simulation services; explicit communication transfer (ECT); protocol specification
Emergency services	TS 102 424	Requirements of the NGN networks to support emergency communication from citizen to authority
	TS 102 164	Emergency location protocols
TISPAN adaptations to 3GPP IMS	TS 181 010	Service requirements for end-to-end session control in multimedia network
	ES 282 007	IP multimedia subsystem (IMS); functional architecture
	TS 182 006	IP multimedia subsystem (IMS); stage 2 description; TS 23 228 Release 6, modified
	TR 183 013	Analysis of relevant 3GPP IMS specifications for use in TISPAN_NGN Release 1 specifications
	ES 283 003	SIP and SDP Stage 3 protocol specification
	TS 183 033	Endorsement of 3GPP TS 29.228 (Release 6) and TS 29 229 (release 6)
	ES 283 027	Endorsement of the SIP-ISUP interworking between the IP multimedia (IM) core network (CN) subsystem and circuit switched (CS) networks
	ES 282 010	IP multimedia subsystem (IMS); Stage 2 description; charging specification
	ES 283 031	IP multimedia; H.248 profile for controlling multimedia resource function processors in the IP multimedia subsystem; protocol specification
PSTN/ISDN emulation subsystem	EN 383 001	Interworking for SIP/SIP-T (BICC, ISUP) [ITU-T Recommendation Q.1912.5, modified]
Softswitch approach	ES 282 002	NGN and PSTN/ISDN emulation (stage 2 of softswitch-based PES)
	TR 183 011	PSTN/ISDN emulation; development and verification of PSTN/ISDN emulation
	ES 283 024	PSTN/ISDN emulation; H.248 profile for controlling trunking media gateways in the PSTN/ISDN emulation subsystem; protocol specification

	ES 283 002	H.248 profile or controlling access and residential gateways in the PSTN/ISDN emulation subsystem; protocol specification
IMS-based approach	TS 182 012	IMS-based PSTN7ISDN Emulation subsystem; Functional Architecture
	TS 183 043	IMS-based PSTN7ISDN Emulation Call Control Protocols; Stage 3
RACS (resource and admission control subsystem)	ES 282 003	Resource and admission control subsystem (RACS); functional architecture
	TS 183 017	Resource and admission control; DIAMETER protocol for session based policy set-up information exchange between the application function (AF) and the service policy decision function (SPDF); protocol specification
	ES 283 026	Resource and admission control; protocol for QoS resource reservation information exchange between the service policy decision function (SPDF) and the access resource control function (A-RACF) in the resource and admission control subsystem; protocol specification
	ES 283 018	Resource and Admission control; H.248 profile for controlling border gateway function (BGF) in the resource and admission control subsystem; protocol specification
	TS 183 017	Resource and admission control; DIAMETER protocol for session based policy setup information exchange between the application function (AF) and the service policy decision function (SPDDF); protocol specification
	ES 283 012	Interworking; trunking gateway control procedures for interworking between NGN and External CD networks
	TS 183 021	Endorsement of TS 29 162 Interworking between IMS and IP networks

Table 8.3 (Continued)

Group name	Number	Title
NASS (network attachment subsystem)	ES 282 004	Network attachment subsystem (NASS); functional architecture
	ES 283 034	Network attachment; DIAMETER based protocol for IP-connectivity related session data exchange between the connectivity session location and repository function (CLF) in NASS and the access-resource and admission control function (A-RASF) in RACS; protocol specification
	ES 283 035	Network attachment subsystem e2 interface based on the DIAMETER protocol
Network Access	TS 183 019	Interface Protocol definitions for network access through xDSL and WLAN access networks
	TS 183 020	Roaming interface protocol definitions for TISPAN NGN network access
Security	TS 187 001	TISPAN NGN security (NGN:SEC); requirements for Release 1
	TR 187 002	Threat and risk analysis
	TS 187 003	Security architecture
NGN overload and congestion control	TR 283 039-3	Overload and congestion control for H."ç((between media gateways and media gateway controllers); protocol specification
Quality of service	TR 102 479	Review of available material on QoS requirements of multimedia services
Network management	TR 188 004	Network management; operations support system; vision
	TS 188 003	Network management; operations support system; requirements
	TS 188 001	Network management; operations support system; architecture
	TR 102 647	Network management; operations support system; standard overview and gap analysis

The TISPAN-NGN R2 will include:

*   evolution of RACS (resource control in the core; towards end-to-end QoS);
*   evolution of NASS (additional access technologies beyond DSL, e.g. WLAN via xDSL, WiMAX etc.);
*   IPTV support;
*   on-line charging;
*   overload control (GOCAP);
*   additional support for mobility and nomadicity;
*   corporate users specific requirements.

8.2.4   TISPAN-NGN Release 3

The TISPAN-NGN R3 will work on generalized mobility, that is, roaming and handover across different access technologies.

8.3   ATIS AND NGN

ATIS (Alliance for Telecommunications Industry Solutions, http://www.atis. org) is an US-based organization. Its mission is to rapidly develop and promote technical and operations standards for the communications and related information technologies industry worldwide using a pragmatic, flexible and open approach. ATIS contributes to the ITU standardization. For NGN, ATIS has produced three very valuable documents for the ITU NGN standards:

*   ATIS NGN Framework Part I: 'NGN Definitions, Requirements, and Architecture'. The document outlines both the network architecture and the technical and business requirements for NGN.
*   ATIS NGN Framework Part II: 'NGN Roadmap'. The document identifies and prioritizes the network architecture capabilities that will enable the introduction of new NGN services. The network architecture capabilities or 'service enablers' include: security, QoS, service decoupling, unified user profile, presence, service transparency, resource and admission control, settlement, NGN management (OAM&P), location-based services, multicast, address resolution, DRM (digital rights management), user control of profile/services, media resource functions and group management.
*   ATIS NGN Framework Part III: 'Standards Gap Analysis'. The document provides an assessment of existing and anticipated NGN standardization activities within the communications industry's standards development community, and identifies areas where standards work is needed.

The ATIS NGN documents are available for members to download.

8.4   CJA AND NGN

CJK (China Japan Korea, http://www.cjk.com) is a standardization body made up of four standards organizations: from China, CCSA; from Japan, ARIB and TTC; and from Korea, TTA. The MoU was signed on 7 November 2002. The objectives of CJK are:

- to exchange views and information on the status of IT industries;
- to monitor the developments of standards issues in IT fields within three countries;
- to encourage mutual support and assistance;
- to contribute to the regional and global standardization bodies.

Its NGN Collaboration Working Group is the one contributing to the ITU NGN standardization. It has:

- cooperated to accelerate the NGN steps in ITU-T, not only on technologies but also in the scope of GSI working areas;
- established the editing team to determine the candidate documents;
- determined the test bed plans.

8.5   TMF AND NGOSS

TMF (Telecom Management Forum, http://www.tmforum.org) is an industry standardization body. Its mission is to provide strategic guidance and practical solutions to improve the management and operation of information and communications services. TMF contributes to the ITU standardization. TMF has been working on the concept and tools for NGN management under the name of NGOSS. TMF claims that NGOSS is the industry's only true standard for development and deployment of easy-to-integrate, flexible, easy-to-manage OSS/BSS components. NGOSS is targeted to define an end-to-end management system for NGN.

Part of the NGOSS has been adopted by ITU and 3GPP already, e.g. NGOSS eTOM levels 1 and 2 for ITU-T and GB923 service quality management for 3GPP. NGOSS is provided as a set of documents that make up a toolkit of industry-agreed specifications and guidelines that cover key business and technical areas, and a defined methodology for use of the tools.

8.5.1   NGOSS Concept [Reproduced with the Permission of TMF]

NGOSS is about designing the NGN architecture and managing NGN, realizing the NGN-required integrated OSS and BSS for the customer-centred operation model. NGOSS defines the framework:

- to design next-generation OSS/BSS to provide an end-to-end management system for the next-generation network, service and support;
- to guide the implementation of business behaviour,
  - with specified key characteristics of OSS/BSS to allow high degrees of process integration and automation;
  - by defining methodologies for evolving OSS and BSS infrastructure into a lean operations approach.

NGOSS identifies and defines a NGOSS lifecycle process/architecture/ methodology, as shown in Figure 8.11, in the sequence:

- Define the business (in business contract).
- Design the architecture for the business (in the system contract).
- Implement the business (implementation of the contract).
- Execute the business (deployment of the contract).

It should also be noticed that:

- the upper part of the business system is technology-agnostic;
- the lower part of the implementation and deployment is dependent on the technology;
- the left part of the business and deployment is relevant for a service provider;
- the right part of system and implementation is relevant for a service developer.

The NGOSS lifecycle provides a framework for the definition, design and development of an NGOSS solution.

8.5.2   NGOSS Components and their Functionality

The NGOSS is composed of four key components, as shown in Figure 8.12, which are: eTOM, SID, TNA&CI and compliance test. The four NGOSS

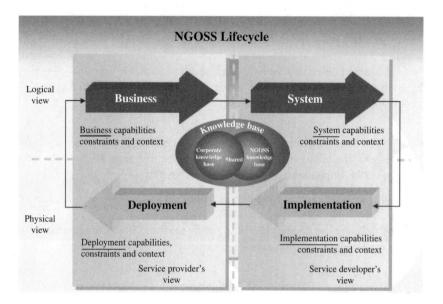

Figure 8.11   NGOSS lifecycle (Reproduced by permission of the TeleManagement Forum)

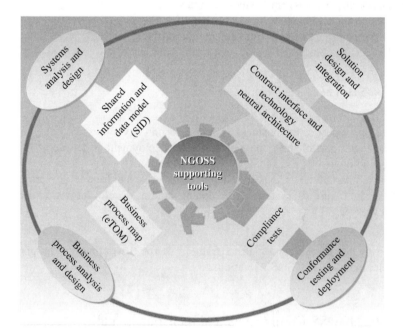

Figure 8.12   TMF NGOSS component view (Reproduced by permission of the Tele-Management Forum)

components fit together to provide an end-to-end system for OSS/BSS development, integration and operations. The NGOSS components may be used as an end-to-end system to undertake large-scale development and integration projects, or used separately to solve specific problems.

The NGOSS components and their functionality are explained below:

- eTOM (enhanced telecom operations map) – defines the next-generation business process framework and creates a common language and a reference view for telecom business processes. It can be used to inventory existing processes at a service provider, act as a framework for defining the scope of a software-based solution, or simply enable better lines of communication between a service provider and their system integrator.
- SID (shared information/data model) – defines systems, people and interactions needed to execute the business processed defined in the eTOM, or a 'common language' for software providers and integrators to use in describing management information, which will in turn allow easier and more effective integration across OSS/BSS software applications provided by multiple vendors. It provides the concepts and principles needed to defined a shared information model, the elements or entities of the model, the business-oriented UML class models, as well as design-oriented UML class models and sequence diagrams to provide a system view of the information and data.
- TNA and CI (technology neutral architecture and contract interface) – create a structure for NGOSS-based solutions using patterns that emphasize reuse of knowledge. It packages them as artefacts and makes them available as NGOSS tools, and makes up the heart of the NGOSS integration framework. In order to successfully integrate applications provided by multiple software vendors, the 'plumbing' of the system must be common. The TNA defines architectural principles to guide OSS developers to create OSS components that operate successfully in a distributed environment. The CI defines the API (application programming interface) for interfacing those elements to each other across the architecture. This architecture is technology-neutral as it does not define how to implement the architecture, rather what principles must be applied for a particular technology-specific architecture to be NGOSS compliant.
- Compliance test – defines tests used to certify software solutions as NGOSS compliant. In order to improve the probability that OSS components will truly integrate with each other, NGOSS provides a suite of tests for compliance with the eTOM, SID, architecture and contract interface components. Any or all of these components can be achieved either singly, or in combination with other components.

8.5.3   NGOSS Documents

The TMF latest delivery of NGOSS is so-called NGOSS Release 6.1. The titles
and summaries of NGOSS Release 6.1 documents are given in the Table 8.4.
The previous NGOSS documents include NGOSS Release 6.0, NGOSS Release
5.0, NGOSS Release 4.5, NGOSS Release 4.0, NGOSS Release 3.5 and
NGOSS Release 3.0. Their titles and summaries are given in Table 8.5. The
NGOSS documents are available for members to download.

Table 8.4   NGOSS Release 6.1 document titles and summaries

Document title	Summary
NGOSS R6.1 Solution Suite Release Notes	This document outlines the changes made in Release 6.1 of the NGOSS Solution Suite
GB922 R6.1 UML Model Files	This document describes the SID Business view (GB926 for SID Systems view) together with a large UML model
NGOSS R6.1 SID Model Suite	This SID document addresses the information and communication service industry's need for shared information/data definitions and models
GB921 Model R6.0 eTOM Browsable and Interactive Version	This document can be used for development and/or browsing within a process tool environment
NGOSS R6.0 Lifecycle Methodology Suite	This document provides the industry with a common framework on how to use and deploy NGOSS within an organization
NGOSS R6.0 eTOM Solution Suite	GB921 Release 6.0 is part of NGOSS 6.0. It updates detailed process decompositions, and an updated and expanded interim release of information on linking eTOM and ITIL
NGOSS R6.0 Compliance Solution Suite	This suite consists of a set of documents that define the strategy for NGOSS testing and define specific tests for compliance with certain aspects of the NGOSS architecture, etc.
NGOSS R6.0 Architecture Solution Suite	This version of the NGOSS contract specification recognizes the role of the contract in an NGOSS system as the fundamental unit of interoperability
NGOSS R6.0 CIM-SID Solution Suite	The Distributed Management Task Force (DMTF) and the TeleManagement Forum (TMF) have each developed a technology-neutral management model of physical and logical devices
GB929 R1.0 Telecom Application Map	This document provides the global telecom software industry with a frame of reference to understand the relationship of the multitude of operational systems typically found within a service provider
NGOSS R4.5 Service Framework Suite	This document provides a recommendation on how a service is modelled into its hierarchy. It outlines the changes made in Release 6.1 of the NGOSS Solution Suite

Table 8.5 NGOSS Releases 6.0, 5.0, 4.5, 4.0, 3.5 and 3.0 document titles and summaries

Document title	Document Summary
NGOSS Release 6.0	Overview and details on NGOSS Release 6.0.
NGOSS 5.0 Overview and Release Notes	In Release 5.0, the NGOSS suite of documents has been enhanced by the addition of a fourth key framework, a Telecom Application Map (TAM)
NGOSS Release 4.5 Overview	Release 4.5 of the TM Forum NGOSS Programme introduces several new items into the NGOSS family of documents. Most notably, the NGOSS suite of documents has been expanded to include the service framework
NGOSS Lifecycle Methodology Suite Release 4.5	The NGOSS Lifecycle project was established to provide the industry with a common framework on how to use and deploy NGOSS within an organization
NGOSS Release 4.5 SID Model Suite (Phase 5)	SID addresses the information and communication service industry's need for shared information/data definitions and models. The definitions focus on business entity definitions and attribute definitions
NGOSS Service Framework Suite Release 4.5	This document provides a recommendation on how a service is modelled into its hierarchical constituent components. The methodology used acknowledges the different functions within a service provider
NGOSS Release 4.0 Overview	The NGOSS Programme of the TMF is driving the direction of OSS/BSS systems for the telecom industry
NGOSS Release 4.0 SID Model Suite (Phase 4)	Phase 4 of the SID Document Solution Suite consists of a set of documents that describe and define the NGOSS SID model. The main entities of the NGOSS business view are substantially completed
NGOSS Architecture Solution Suite Release 4.0	The TMF053 NGOSS Architecture series of documents describe the major concepts and architectural details of the NGOSS architecture in a technologically neutral manner
NGOSS Compliance Solution Suite Release 4.0	The NGOSS Compliance Solution Suite consists of a set of documents that define the strategy for NGOSS testing and define specific tests for compliance to certain aspects of the NGOSS architecture
NGOSS Lifecycle Methodology – GB927 V1.1	The NGOSS Lifecycle project was established to provide the industry with a common framework on how to use and deploy NGOSS within an organization. It covers identifying and describing a business

8.6 NGMN ALLIANCE AND NGMN, AND 3GPP AND LTE/SAE

8.6.1 NGMN Alliance and NGMN

NGMN Alliance (Next Generation Mobile Network Alliance, formerly NGMN Initiative, http://www.ngmn.org) was founded in 2006 by leading global mobile operators. Today the Alliance has been extended to 13 mobile operators as members, 17 technology vendors as sponsors and two universities as advisors, where:

- the NGMN member operators are Alltel, AT&T, China Mobile, KPN Mobile, NTT DoCoMo, Orange, SK telecom, Sprint, T-Mobile, Telecom Italy, Telefonica, TELUS and Vodafone Group;
- the NGMN sponsor vendors are Airvana, Huawei, LG, NEC, Samsung, Starent, ZTE, Alcatel-Lucent, Cisco, Ericsson, Intel, Motorola, Nokia, Noki-aSiemensNetworks, Nortel, Qualcomm, and Texas Instruments;
- the two NGMN advisors are the University of Surrey and the UMIC RWTH-Aachen University.

The NGMN Alliance has provided a vision for technology evolution beyond 3G for the competitive delivery of mobile broadband services to further increase end-customer benefits. The objective is to establish clear performance targets, fundamental recommendations and deployment scenarios for a future wide area mobile broadband network, and to make sure that its price/performance is competitive with alternative technologies. This initiative intends to complement and support the work within standardization bodies by providing a coherent view of what the operators community is going to require in the decade beyond 2010.

Emphasis is also on the IPR side, where the goal is 'to adapt the existing IPR regime to provide a better predictability of the IPR licenses to ensure Fair, Reasonable and Non-Discriminatory (FRAND) IPR costs' [31]. The outcome recommendations have been summarized in the white paper entitled 'Next Generation Mobile Networks beyond HSPA and EVDO'. The latest version of version 3.0 was published in December 2006, and is available at the home page of NGMN Alliance.

The targeted timelines are to enable commercial services on a country- and operator-specific basis by 2010, assuming standards to be completed by the end of 2008 and for operator trials to be supported in 2009 along with the availability of mobile devices in sufficient volumes and at a sufficient quality level at the same time.

In the white paper version 3.0, the NGMN initiative has accepted, as working assumptions, that:

- 3GPP LTE will be one of the most likely vehicles for the delivery of the NGMN radio design.
- 3GPP SAE is a prime candidate for the delivery of the NGMN system architecture and, as such, it must meet the NGMN architecture requirements stated in the white paper.

8.6.2   3GPP and LTE/SAE

3GPP, the 3rd Generation Partnership Project (http://www.3gpp.org), was created in December 1998 by ETSI and a number of regional partners and is the dedicated standardization body for UMTS specification and GSM evolution. 3GPP is supported by the MCC (Mobile Competence Centre) hosted by ETSI. LTE/SAE is an 3GPP concept, a long-term evolution for the 3GPP radio access technology and core network, where:

- LTE is an advanced radio access technology, for the enhancement of UTRAN;
- SAE is an advanced core network, for the enhancement of PS technology over IP CN.

LTE and SAE are being approached independently, but they enhance each other and have become inseparable. An accepted industry agenda for LTE/SAE development and deployment is given in Figure 8.13.

3GPP LTE (Long Term Evolution) [Reproduced with the permission of ETSI] [32,33,35]

In November 2004 in Toronto, Canada, a 3GPP RAN Evolution Workshop dedicated to the Evolution of the 3G Mobile System marked the start of the work on 3GPP LTE. During this workshop, operators, manufacturers and research institutes presented more than 40 contributions with views and proposals on the evolution of UTRAN (Universal Terrestrial Radio Access Network).

Figure 8.13   Industry agenda for LTE/SAE development and deployment

Among others, a set of high-level requirements on the UTRAN evolution was identified:

- reduced cost per bit;
- increased service provisioning – more services at lower cost with better user experience;
- flexibility of use of existing and new frequency bands;
- simplified architecture, open interfaces;
- allow for reasonable terminal power consumption.

In December 2004, with the conclusions of this workshop and with broad support from 3GPP members, a feasibility study on the LTE of UTRA and UTRAN was started. The objective was 'to develop a framework for the evolution of the 3GPP radio-access technology towards a high-data-rate, low-latency and packet-optimised radio-access technology'.

The study focused on supporting services provided from the PS-domain, involving the following items:

- related to the radio-interface physical layer (downlink and uplink), e.g. it must have the means to support flexible transmission bandwidth up to 20 MHz, the introduction of new transmission schemes and advanced multi-antenna technologies;
- related to radio interface layers 2 and 3, e.g. signalling optimization;
- related to the UTRAN architecture and identifying the optimum UTRAN network architecture and functional split between RAN network nodes;
- RF-related issues.

In addition, the 3GPP has also taken the set of recommendations from the NGMN Initiative into consideration for the creation of networks suitable for the competitive delivery of mobile broadband services. The recommendations include the key system characteristics and the detailed requirements. As a study result, the detailed requirements of LTE have been worked out for the following criteria:

- Peak data rate
  - Instantaneous downlink peak data rate of 100 Mbps within a 20 MHz downlink spectrum allocation (5 bps/Hz), possibly will be upgraded to >300 Mbps.
  - Instantaneous uplink peak data rate of 50 Mbps (2.5 bps/Hz) within a 20 MHz uplink spectrum allocation, possibly will be upgraded to >80 Mbps.

- Control-plane latency

  - Transition time of less than 100 ms from a camped state, such as Release 6 Idle Mode, to an active state such as Release 6 CELL_DCH.
  - Transition time of less than 50 ms between a dormant state such as Release 6 CELL_PCH and an active state such as Release 6 CELL_DCH.

- Control-plane capacity

  - At least 200 users per cell should be supported in the active state for spectrum allocations up to 5 MHz.

- User-plane latency

  - Less than 5 ms in the unloaded condition (i.e. single user with single data stream) for small IP packets.

- User throughput

  - Downlink: average user throughput per MHz, three to four times Release 6 HSDPA.
  - Uplink: average user throughput per MHz, two to three times Release 6 Enhanced Uplink.

- Spectrum efficiency

  - Downlink: in a loaded network, the target for spectrum efficiency (bps/Hz/site) is three to four times Release 6 HSDPA.
  - Uplink: in a loaded network, the target for spectrum efficiency (bps/Hz/site)is two to three times Release 6 Enhanced Uplink.

- Mobility

  - E-UTRAN should be optimized for low mobile speed from 0 to 15 km/h.
  - Higher mobile speed between 15 and 120 km/h should be supported with high performance.
  - Mobility across the cellular network should be maintained at speeds from 120 to 350 km/h (or even up to 500 km/h depending on the frequency band).

- Coverage

  - Throughput, spectrum efficiency and mobility targets above should be met for 5 km cells, and with a slight degradation for 30 km cells. Cells ranging up to 100 km should not be precluded.

- Further enhanced multimedia broadcast multicast service (MBMS)

  - While reducing terminal complexity, the same modulation, coding, multiple access approaches and UE bandwidth as for unicast operation.
  - Provision of simultaneous dedicated voice and MBMS services to the user.
  - Available for paired and unpaired spectrum arrangements.

- Spectrum flexibility

  - E-UTRA should operate in spectrum allocations of different sizes, including 1.4, 1.6, 3, 3.2, 5, 10, 15 and 20 MHz in both the uplink and downlink. Operation in paired and unpaired spectra should be supported.
  - The system should be able to support content delivery over an aggregation of resources including radio band resources (as well as power, adaptive scheduling, etc.) in the same and different bands, in both uplink and downlink and in both adjacent and non-adjacent channel arrangements. A 'radio band resource' is defined as all spectra available to an operator.

- Co-existence and interworking with 3GPP radio access technology (RAT)

  - Co-existence in the same geographical area and co-location with GERAN/UTRAN on adjacent channels.
  - E-UTRAN terminals also supporting UTRAN and/or GERAN operation should be able to support measurement of, and handover from and to, both 3GPP UTRAN and 3GPP GERAN.
  - The interruption time during a handover of real-time services between E-UTRAN and UTRAN (or GERAN) should be less than 300 ms.

- Architecture and migration

  - Single E-UTRAN architecture.
  - The E-UTRAN architecture should be packet-based, although provision should be made to support systems supporting real-time and conversational class traffic.
  - E-UTRAN architecture should minimize the presence of 'single points of failure'.
  - E-UTRAN architecture should support end-to-end QoS.
  - Backhaul communication protocols should be optimized.

- Radio resource management requirements

  - Enhanced support for end-to-end QoS.
  - Efficient support for transmission of higher layers.
  - Support of load sharing and policy management across different radio access technologies.

- Complexity

  - Minimize the number of options.
  - No redundant mandatory features.

The wide set of options initially identified by the early LTE work was narrowed down, in December 2005, to a working assumption that the downlink would use orthogonal frequency division multiplexing (OFDM) and the uplink would use single carrier–frequency division multiple access (SC-FDMA). Although opinions were divided, it was eventually concluded that inter-Node-B macro-diversity would not be employed.

Supported downlink data-modulation schemes are QPSK, 16QAM and 64QAM. The possible uplink data-modulation schemes are (pi/2-shift) BPSK, QPSK, 8PSK and 16QAM.

The use of a multiple input multiple output (MIMO) scheme was agreed, with possibly up to four antennas at the mobile side, and four antennas at the cell side. Re-using the expertise from the UTRAN, the same channel coding type as for UTRAN was agreed (turbo codes). To approach the radio interface protocols of the evolved UTRAN, the initial assumptions were:

- simplification of the protocol architecture and the actual protocols;
- no dedicated channels, and hence a simplified MAC layer (without MAC-d entity);
- avoiding similar functions between radio and core network.

A transmission time interval (TTI) of 1 ms was agreed (to reduce signalling overhead and improve efficiency). RRC states were restricted to RRC_Idle and RRC_Connected states. They are depicted in Figure 8.14, in conjunction with the possible legacy UTRAN RRC states [32].

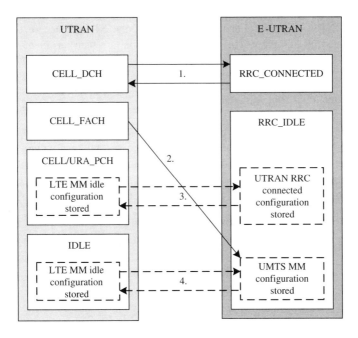

Figure 8.14   RRC state (Reproduced from http://www.3gpp.org/Highlights/LTE/LTE.htm)

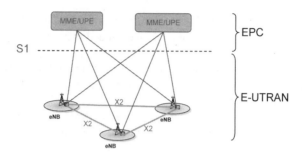

Figure 8.15 Evolved UTRAN architecture (Reproduced from http://www.3gpp.org/ Highlights/LTE/LTE.htm)

The evolved UTRAN architecture is given in Figure 8.15 [34], where:

- The eUTRAN consists of eNBs, providing the evolved UTRA U-plane and C-plane protocol terminations towards the UE.
- The eNBs are interconnected with each other by means of the *X2 interface*. It is assumed that there always exists an X2 interface between the eNBs that need to communicate with each other, e.g. for support of handover of UEs in LTE_ACTIVE.
- The eNBs are also connected by means of the S1 interface to the EPC (evolved packet core). The *S1 interface* supports a many-to-many relation between aGWs and eNBs.

The functional split of network entities is:

- The eNB hosts the functions for radio resource management including radio bearer control, radio admission control, connection mobility control and dynamic resource allocation (scheduling).
- The mobility management entity (MME) is for the distribution of paging messages to the eNBs.
- The user plane entity (UPE) is for:
  - IP header compression and encryption of user data streams;
  - termination of U-plane packets for paging reasons;
  - switching of U-plane for support of UE mobility.

To cooperate on this functional split, the protocol stack is designed as in Figure 8.16 [33].

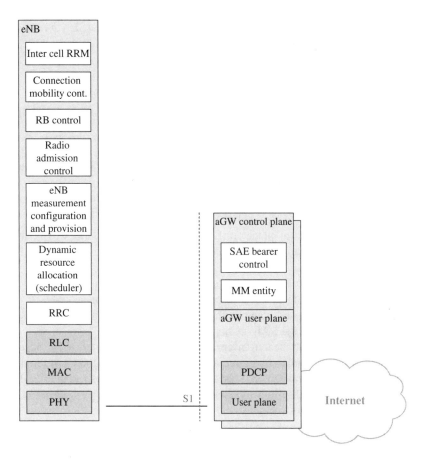

Figure 8.16 Protocol stack of evolved UTRAN architecture (Reproduced from http://www.3gpp.org/Highlights/LTE/LTE.htm)

3GPP SAE (System Architecture Evolution) [9]

Separately from LTE, 3GPP started studying the system architecture evolution (SAE) with the objective 'to develop a framework for an evolution or migration of the 3GPP system to a higher-data-rate, lower-latency, packet-optimised system that supports multiple RATs. The focus of this work is on the Packet Switched (PS) domain with the assumption that voice services are supported in this domain'.

In December 2004, the 3GPP SAE work was approved. It was initiated when it became clear that the future was clearly IP with everything, and that access to the 3GPP network would ultimately be not only via UTRAN or GERAN but also by WiFi, WiMAX, or even wired technologies. Thus SEA has as its main objectives:

- impact on overall architecture resulting from LTE work;
- impact on overall architecture resulting from AIPN work;
- overall architectural aspects resulting from the need to support mobility between heterogeneous access networks.

The 3GP SAE is shown in Figure 8.17. The new reference points are defined as:

- S1 provides access to evolved RAN radio resources for the transport of user plane and control plane traffic. The S1 reference point will enable MME and UPE separation and deployments of a combined MME and UPE solution.
- S2a provides the user plane with related control and mobility support between a trusted non-3GPP IP access and the SAE anchor.
- S2b provides the user plane with related control and mobility support between ePDG and the SAE anchor.
- S3 enables user and bearer information exchange for inter-3GPP access system mobility in idle and/or active states. It is based on the Gn reference point as defined between SGSNs. User data is forwarded for inter-3GPP access system mobility in active state (FFS).

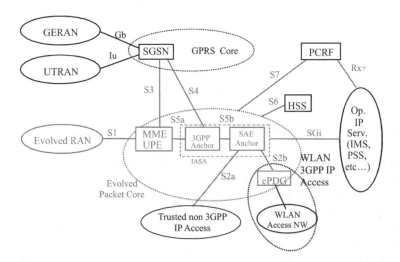

Figure 8.17 3GPP SAE architecture (Reproduced from http://www.3gpp.org/ Highlights/LTE/LTE.htm)

- S4 provides the user plane with related control and mobility support between GPRS core and the 3GPP anchor and is based on the Gn reference point as defined between SGSN and GGSN.
- S5a provides the user plane with related control and mobility support between MME/UPE and 3GPP anchor. It is FFS whether a standardized S5a exists or whether MME/UPE and 3GPP anchor are combined into one entity.
- S5b provides the user plane with related control and mobility support between 3GPP anchor and the SAE anchor. It is FFS whether a standardized S5b exists or whether 3GPP anchor and SAE anchor are combined into one entity.
- S6 enables transfer of subscription and authentication data for authenticating/authorizing user access to the evolved system (AAA interface).
- S7 provides transfer of (QoS) policy and charging rules from PCRF to the policy and charging enforcement point (PCEP).
- The allocation of the PCEP is FFS.
- SGi is the reference point between the inter-AS anchor and the packet data network. The packet data network may be an operator-external public or private packet data network or an intra-operator packet data network, e.g. for provision of IMS services. This reference point corresponds to Gi and Wi functionalities and supports any 3GPP and non-3GPP access systems.
- The interfaces between the SGSN in 2G/3G core network and the EPC will be based on the GTP protocol. The interfaces between the SAE MME/UPE and the 2G/3G core network will be based on the GTP protocol.

REFERENCES

[1] ITU-T Y.2011, 'General principles and general reference model for next-generation networks'.
[2] ITU-T M.3060, 'Principles for the management of next generation networks'.
[3] ITU-T NGN – GSI Release 1 'NGN_FG-book_II'.
[4] ITU-T Y.2021: 'IMS for next generation networks'.
[5] ITU-T Q.1236: 'Intelligent network capability set 3 – management information model requirements and methodology series Q: switching and signalling intelligent network'.
[6] JT Q.1200: 'Guidance for JT-Q1200 series standards (intelligent network)'.
[7] 3GPP TS 22.228: 'Service requirements for the Internet protocol (IP) multimedia core network subsystem'.
[8] TIA-771: 'Wireless intelligent network'.
[9] 3GPP TR 23.882: '3GPP system architecture evolution, report on technical options and conclusions'.
[10] OSA-Parlay-4.
[11] OSA-Parlay-5.
[12] OSA-Parlay-X.
[13] OMA-OSE.

[14] ITU-T Y2012: 'Functional requirements and architecture of the NGN'.

[15] ITU-T G.992.1.

[16] ITU-T G.992.3.

[17] ITU-T G.992.5.

[18] ITU-T G.991.2.

[19] ITU-T G.993.1.

[20] ITU-T G.993.2.

[21] ITU-T G.707: 'Network node interface for the synchronous digital hierarchy (SDH)'.

[22] IEEE 802.3ah (100Base-LX/BX).

[23] ITU-T G.983.x series for B-PON (Broadband- Passive Optical Network).

[24] ITU-T G.984.x series for G-PON (Gigabit – Passive Optical Network).

[25] IEEE 802.3ah (1000Base-PX).

[26] ITU-T J.179 'IPCablecom support for multimedia'.

[27] IEEE 802.3.

[28] IEEE 802.3u.

[29] IEEE 802.3z.

[30] IEEE 802.3ae.

[31] NGMN White paper version 3.0: 'Next generation mobile networks beyond HSPA&EVDO', December 2006.

[32] 3GPP TR 25.814: 'Physical layer aspects for evolved universal terrestrial radio access (UTRA)'.

[33] 3GPP TR 25.913: 'Requirements for evolved UTRA (E-UTRA) and evolved UTRAN (E- UTRAN)'.

[34] ETSI ES 282 001 version 1.1.1: 'Protocols for advanced networking (TISPAN); NGN Functional Architecture Release 1'.

[35] ETSI: 'Long term evolution of the 3GPP radio technology' and 'System architecture evolution'.

9

NGNs and Corporate Responsibility

The information and communication technologies already ease the life of millions and bring many advantages to the users in terms of access to information and knowledge. Stepping toward NGN, this sector has the potential to bring much more, so much in fact, that ICT have been integrated in many national and international strategies for sustainable development as one of its fundamental elements. The European Union, in its Lisbon Strategy launched in 2000, included ICT as a key element in its strategy for sustainable development.

However, like so many good things, ICT unfortunately comes with some negative aspects that need to be clearly identified. Clarity about the negative as well as about the positive aspects of products and services, the proper behaviour in handling these aspects and undertaking the correct actions are the essence of corporate responsibility.

In parallel to this, there are also nowadays strong signals towards a return of business ethics, due greatly to business and accounting scandals that shook the world at the turn of the century, driving law enforcement officers inside companies and investors away. This is not so new; every wave of business scandals has been followed by a wave of moralization, as investors painfully learn to distinguish between responsible and irresponsible management.

This time though, it is slightly different: besides financial issues, other key issues are starting to markedly shape the way companies conduct business. Environmental and social aspects are of course in the first line, but due to globalization, the technological consequences of science, with their universal reach, require an ethical treatment from companies.

In this chapter, the concept of sustainable development is introduced and then its link to corporate responsibility will be shown. We will then describe what

Next Generation Networks: Perspectives and Potentials   Jingming Li Salina and Pascal Salina
© 2007 John Wiley & Sons, Ltd

the responsibility of companies covers and why it belongs today to the scope of modern management. The specific aspects of corporate responsibility and sustainable development in the field of ICT and NGN are finally highlighted.

9.1   UNSUSTAINABLE GROWTH

Economic growth is constantly advocated, but it is unfortunately no longer sustainable under the present conditions. Even the most obtuse person cannot deny today the over-consumption of resources in our free economy societies.

Visions of the future are powerful growth drivers, especially when these visions are easy to understand: 'a Coke available to everyone everywhere in the world' or 'a computer in every home', as simple they may sound, truly drove two outstanding companies towards new levels of achievements. Piggy-backing on the demographics, these visions were deemed to succeed.

This will not be the case in the future as our ecological and social footprints have become so large that we will need more than one planet just to survive! Reality demands that this issue be tackled by designing scenarios to reduce our ecological footprint. This is shown in a striking manner in Figure 9.1.

The human ecological footprint (HEF) describes the amount of land and area needed by a human population to produce resources and absorb waste or emissions. The approximate HEF is based on only three components: agricultural land, urban-industrial land and CO_2-absorption land. Modelling the HEF shows through simulation that one planet Earth is not any longer sufficient to sustain the comfort and resource consumption of technologically developed societies

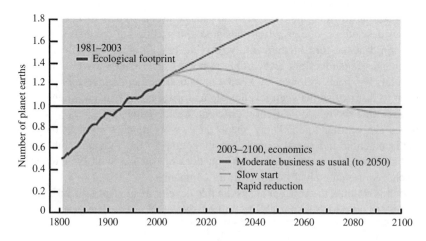

Figure 9.1   Human ecological footprint (Reproduced by permission of the WWF – World Wildlife Fund for Nature, formerly World Wildlife Fund)

(source: WWF [1], using the World3 System Dynamic Computer Model; other models of the carrying capacity of the earth exist [2]).

Exponential human population growth, climate change, depletion of non-renewable resources, extreme strains or even destruction of ecosystems and its corollary, the drastic reduction of biological diversity [3], are today's ecological and social mega-trends that cannot be ignored any more when setting targets and designing the relevant corporate strategies. These trends are not cyclical on a human time scale. Worse, they appear likely to be irreversible; hence the importance of taking the right course into the future as one cannot wait for the return of better times. Companies must integrate this new paradigm, but how?

9.2 SUSTAINABLE DEVELOPMENT AND CORPORATE RESPONSIBILITY

When faced with the task of solving the dilemma of economic growth as a way to reduce poverty and inequalities without destroying the environment for future generations, the World Commission on Environment and Development, headed by Mrs Gro Harlem Brundtland, in its final report spoke of 'our common future' [4] and the concept of 'sustainable development'.

This concept describes pathways to achieving an equitable, liveable and safe future, 'improving the quality of life while living within the carrying capacity of supporting ecosystems' (UICN, UNEP and WWF, 1991).

Characterization of a development as sustainable:

- implies that its benefits are maintained indefinitely;
- requires, by definition, that the future be determined, which therefore brings a degree of uncertainty.

Strategies and measures selected must fit these two requirements. There cannot be overall any activities that would be damaging or reduce overall value. Regarding the second point, the surest way to alleviate the uncertainty is to choose and follow those actions that experience has shown to be sustainable and reject those that are clearly unsustainable. As presented by the Brundtland Commission though, this includes the satisfaction of today's needs as well as and without limiting the satisfaction of the needs of future generations. This definition explicitly includes a strong time dimension since it postulates an arbitrage between the satisfaction of present and future needs.

There is a 'principle of responsibility' at work here, an old philosophical principle of ethics, which not only requires one to be accountable for one's actions in the present and immediate future in the classical way, but also requires one to consider his or her impact in the long term or the very long term (many generations).

Briefly stated, the principle of responsibility stipulates that we are responsible for the future we prepare and build for many generations. It is not the authority of the past that should orientate people any longer, but their projects for the future: 'the wise man must remember that while he is a descendant of the past, he is a parent of the future' (Herbert Spencer) or with a lighter touch: 'My interest is in the future because I'm going to spend the rest of my life there' (attributed to Charles Kettering).

As companies will be accountable to future generations, it is therefore their responsibility to follow the concept of sustainable development in order to resolve, or at least as a first step in resolving the dilemma of growth and the preservation of the environment for future generations. In this manner, sustainable thinking meets the interests of both the shareholders and the stakeholders.

Nevertheless, the concept of sustainable development as coined by the Brundtland Commission is difficult to use in everyday corporate life. The question of arbitrage over time, although central to sustainable development, is most of the time removed from the reflections of corporations and conveniently left to entities not forced by the stock market to present quarterly results. The prediction of the future (the time factor) introduces a degree of uncertainty.

With one key dimension slightly shadowed, the concept was thus simplified to cover only the social, ecological and economical aspects in the present or short term. It was also made somewhat more understandable by such popular expression as the 'triple P' slogan, standing for People, Planet and Profit, see Figure 9.2.

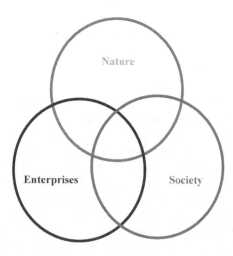

Figure 9.2   The now classical representation of sustainable development and its three dimensions

It is necessary here to mention that no ultimate definition of corporate responsibility exists today and many views on what corporate responsibility is or is not coexist. Sometimes still called CSR or corporate social responsibility, it should be understood as the responsibility of management to consider and include broader issues such as environmental and social impacts of its activities. This is the only correct route toward sustainable development, itself a possible way out of the contradiction between economic growth and the protection of environment.

To sum up the link between sustainable development and corporate responsibility, we can say that sustainability has become the most important principle of action of a responsible company as it could solve the growth dilemma. A company can be regarded as responsible only if it follows the principle of sustainable development.

9.3   THE PURPOSE OF CORPORATE RESPONSIBILITY

'Companies operate within well-defined legal frameworks and their responsibility is simply to comply with the law'. 'The rest is only a hindrance to business and regulations should be kept to a minimum'. 'Anyway, is there a business case for corporate responsibility?' These kinds of statements can still be heard today across boards and at all management levels. These depressively narrow views are not taking into account the vital interest that the long-term prosperity of a company brings benefits to all and the fact that new relationships are being built with individuals or groups who take the view that we all have something in common at stake in this finite world. These stakeholders are better informed than ever (thanks to ICT) and can get and spread information very quickly in order to defend rights considered legitimate (thanks again to ICT!). They are not only activists; they include prospective customers, employees and suppliers, communities, elected authorities, researchers, trade unions and trade associations, media and opinion leaders to name but a few.

There is another point. It is clear that businesses are neither philanthropic nor peacekeeping organizations. They are driving forces for prosperity. This prosperity rest on products and services created and marketed by companies, as well as on the redistribution of their turnover in salaries, payments, investments and of course dividends.

However, is there really a net positive value created if at the same time natural resources are depleted, eco-systems destroyed, biological and cultural diversity reduced and people constrained? There must be net (positive) benefits in the three dimensions of social, environmental and financial aspects in order to claim value creation. This is what the picturesque expression of 'triple bottom line' [5] means. The accounting must show a positive value 'under the bottom line' when summing up items in each of the three dimensions. No one today will trust a statement of value creation if there is at the same time a destruction of ecological and social values behind the doors.

Then how to reconcile the legitimate profit-making objectives of the business sector and the new imperatives of responsibility? The purpose of corporate responsibility is precisely that.

It is to analyse and to balance the positive and negative aspects of business, products and services. It is the behaviour a corporation should adopt or follow when facing impacts, risks and conflicts of interest, and issues arising from the technology, products or services it brings to the market. It is to assure customers that they can consume products and services whose impacts have been mitigated or even eliminated. It is to demonstrate to shareholders that their assets are not at risk or will not cause impacts that will destroy overall value. It is ultimately to build and to maintain trust and confidence among stakeholders in order to pursue operations in a sustainable manner.

Figure 9.3 shows the ecological, economic and social dimensions of sustainable development. According to the principle of sustainable development, companies taking into account all the dimensions are much better off than companies focusing only on the usual economic dimension, as represented by the length of the diagonal arrow (the sum of the three vectors). If a company does not take sufficient account of the social and ecological aspects of its activities, it could end up in a situation where overall value is actually destroyed: the economic dimension could well grow, but if the two other dimensions are negative, the resulting vector will point backward, a sure sign of an unsustainable situation. Such companies do not contribute to the prosperity of society.

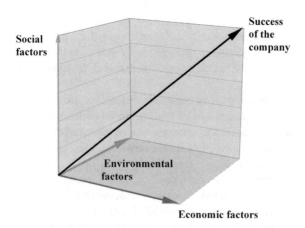

Figure 9.3   A geometric representation of sustainable development

9.4   THE FUNDAMENTALS AND THE LIMITS OF CORPORATE RESPONSIBILITY

9.4.1   Principles and Values

Corporate responsibility derives of course from the legal requirements imposed upon the company, but it goes beyond this. Articles of Incorporation and a vision and mission statement set the frame of economic activities, but values lay the fundamentals for the responsible attitude of a company toward its stakeholders, the society and the environment.

Corporate culture is built upon these fundamental values of a company. Both values and corporate culture are important framing elements and driving forces for a sustainable corporate management as they flow into the guiding policies of the company and in its strategy. Values play an all-important structuring role and ensure coherency and consistent policies and strategies.

A few principles should be considered when setting the values of the company. The principle of responsibility was mentioned in Section 9.2 as perhaps one of the most important. Two other principles revolve around the responsibility principle and help shape the attitude of a responsible company.

The principle of prevention, which requires a focus on preventing negative impacts in opposition to waiting and repairing damage, is a powerful principle of action. It supposes that scientific certainty about the effects exists and that impacts of activities are clearly identified and possibly quantified, which is unfortunately not always the case, especially with new technologies.

In case of uncertainty, a third principle comes into play, typically when introducing a novel technology: the principle of precaution, with the aim of anticipating and minimizing potentially serious risks. This principle is probably the one that has the greatest meaning in determining responsibility. It stipulates that activities shall not be undertaken as long as their potential effects on the environment or health or on the social sphere have not been evaluated, and that the absence of proof of a negative effect is not a sufficient reason to refuse to undertake preventative measures.

A word of caution is necessary on the principle of precaution: by no means is it a reason in itself to avoid a new technology or slow down or forbid its introduction. It is merely providing a framework to evaluate risks as early as possible, thus defining room for improvement and future development of the technology in question.

Debate on the interpretation of the principle of precaution still goes on and is a heated one, as the application (or the non-application) of the principle can have far-reaching consequences for society.

It can be noted with interest that standardization bodies of the ICT sector have started to consider the principle of precaution in the development of new standards. Experts, dedicated task teams in liaison with relevant expert bodies, are increasingly included at early stages of standardization. One example is the consideration of health issues when developing standards on communication technologies using electromagnetic fields.

9.4.2   The Limits of Corporate Responsibility

If corporate responsibility goes beyond mere compliance with legal requirements, how far does it extent? A very simple example will set the problem straight. In 2006, the world's population must share the same constant amount of fresh water [6] as the population of 1876, the year Alexander Graham Bell invented the telephone, assuming for the sake of simplicity that eco-systems and cycles purifying and preparing the water remain untouched and that the stock of fresh water is not diminishing. With a world population of 6 billion in 2006 and supposedly of 1.3 billion in 1876, this represents a decrease of 460 % of fresh water available per capita to satisfy basic human needs, assuming again for simplicity that the need per capita for fresh water has stayed the same.

Companies with water-intensive processes will enter into costly competition for clean water, besides facing the challenge of polluting as little as possible the water they output. These double constraints are already a reality in many countries.

This very simple example can be applied to virtually all natural resources; for instance a bitter controversy erupted at the beginning of 2007 over the use of corn as food in Mexico or as fuel (through fermentation into ethanol) in the USA [7]. We will see more conflicts of this type, opposing a moral issue and an economic one.

In conclusion, simple market mechanisms are no longer adequate to solve the problem of proper allocation of non-renewable resources. Companies cannot wait for the market to react and must become proactive and tackle these problems themselves, for their own sake and for the sake of their customers and shareholders. Their corporate responsibility thus extends beyond the limits of companies, upstream and downstream, and binds the three aspects of sustainable development over time: the economic cost of a resource, its ecological cost and the social cost of its use.

Figure 9.4 illustrates the extent of corporate responsibility to the three domains of sustainable development. Borne by the domain of the 'enterprise', it includes positive and negative impacts and outcomes in the two other spheres of 'nature' and 'society'. The interactions between the sphere delimitate the domains where corporate responsibility is particularly important for companies: those domains cover the modes of behaviour compatible with our environment and our societies where companies may be held accountable for their attitude or where they could proactively grasp the opportunity to build and maintain trust and confidence among customers and stakeholders.

As time is an important dimension of sustainable development, it cannot actually be removed from reflection on corporate responsibility. With the

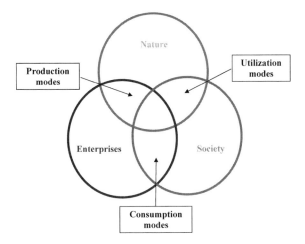

Figure 9.4   The extent of corporate responsibility

emergence of trends that are not cyclical (in the horizon of business, the short and medium term), such as demographics, climate change and the destruction of ecosystems, questions about the future must be addressed.

Paraphrasing or enhancing the essentials of the well-known book *Competing for the Future* [8], responsible, sustainable companies must tackle the following issues:

- How compatible are our production methods with the ecological sphere (see Figure 9.4, production modes)?
- What consumption patterns will we favour among our future customers (see Figure 9.4, consumption modes)?
- What will influence our business and our products; will our products be renewable, refurbishable, recyclable (see Figure 9.4, utilization modes)?
- Which natural resources will we need to operate our business in the future; which eco-system will deliver them?
- What alternative resources will we have, will we need?
- What efficiency will be required with the sparse resources then available?
- What amount of energy will be necessary; how long can we still rely on fossil energy after the oil peak; what type of renewable energy should we choose and will it be impacted by climate change?
- Which external environmental and social impacts will slow us down?
- What skills and competences will be necessary; will the future education system be adequate?

Recognizing the importance of these topics, spotting the threats and the opportunities, and answering these questions, although challenging, is the duty of a responsible management towards present and future customers, shareholders and stakeholders.

Who is ultimately setting the limits of corporate responsibility? This is a delicate matter, which is best solved in common with stakeholders. If the sphere of obligations of a company is set by legal requirements and contracts, its sphere of influence and impacts goes beyond them and its boundaries will be determined together with stakeholders. Tackling direct impacts belongs without discussion to the responsibility of an organization. Beyond this, tackling indirect impacts or impacts that can be moderately influenced, is mostly a purely voluntary engagement and belongs in the realm of truly good corporate citizen; or could it be a subtle method to increase reputation while reducing risks at the same time?

The problem of supply chains is a typical example of issues faced by modern companies: vital bloodstreams of business, supply chains must be carefully organized and maintained. Depending on their length, their origin and their regional or international exposure, they exhibit very different risk profiles and types of challenges and impacts. Setting common standards along the whole chain, beyond the reach of direct control, is probably one the most complex problems for responsible companies.

9.5 STANDARDS AND TOOLS OF CORPORATE RESPONSIBILITY

Although there is yet no standard defining corporate responsibility in a comprehensive way [the ISO 26000 standard, originally on Corporate Social Responsibility (CSR), now renamed Social Responsibility (SR), is due to be published in October 2008], there is a collection of competing normative documents in the social and ecological dimension that cannot be totally ignored. They are issued by widely different bodies, some 'public' (from government and established international bodies), some 'private' (from non-governmental or not-for-profit organizations, trade associations or other interest groups).

This trend toward social and ethical accounting, auditing and reporting is sometimes identified by its acronym SEAAR; activities usually performed and controlled by third parties. This field of activity is now heavily crowded, compliance control and transparency being a very lucrative business. Against the flow, one can support the alternative view that certified management systems (typically environmental management systems according to ISO 14001) render the external control of reporting superfluous.

The whole concept of corporate responsibility tends nowadays to become highly codified but, although quite helpful, the documents and guidelines available and mentioned hereafter are by no means the only way to address corporate responsibility.

Most effort has been made in the field of the environment, a fact reflected in the greater number of norms, standards and guidelines available, resulting from a consensus on most facts and scientific knowledge of natural phenomena. In the social sphere, the standardization of ethical practices is much less advanced, as the complexity and the sensitivity of the topic require great care in tackling the issues.

The most useful or recognized standards and tools are presented here, but the list is not exhaustive.

9.5.1   Norms

Environment

The family of *ISO 14000 documents* standardizes different aspects of environmental management. There are more than 20 standards, guides and other publications dealing with environmental issues, from management systems to assessment methods. The most prominent of the ISO 14000 family are:

- The *ISO 14001:2004* – this international standard normalizes environmental management systems. It integrates a mechanism of continuous improvement to reduce pollution, based on the circular Plan–Do–Check–Act of William Edward Deming. The revised version, published in 2004, requires certified companies to consider not only their direct impacts but also the indirect environmental aspects they can influence, among other things. A certification of management systems following this standard is possible and lasts for a period of three years.
- The *ISO 14030 series* (Environmental Performance Evaluation, EPE) normalize the monitoring of environmental performance of organizations.
- The *ISO 14040 series* (Life Cycle Analysis, LCA) normalizes the assessment of the life cycle of products, an important tool in assessing the environmental impacts along the life cycle of a product, from its production to its disposal or recycling, and in the prioritization of impacts.
- *ISO 14062* (Design for Environment, DfE) deals with the integration of environmental aspects in the design and development of products.
- *ISO 14063* (Environmental Communication) provides a guideline and examples that will help companies to make the important link to external stakeholders.
- *ISO 14064* (Greenhouse Gases) is a guideline on greenhouse gases monitoring and reporting

Other Types of Environmental Standards

The *EMAS* (Eco-management and Audit Scheme) European standard normalizes environmental management systems and their certification. It requires organizations to register on a site-to-site basis (whereas an ISO 14001 certification may cover one system valid for many sites). It is more stringent than ISO 14001, puts more emphasis on environmental reporting and requires a greater participation of employees. Its revision in 2001 (EMASII) integrates the ISO 14001 as the environmental management system. This scheme is valid throughout Europe and EEA countries.

Social or More Socially Oriented Standards

International Labor Organization (ILO) Declaration on Fundamental Principles and Rights at Work: The Declaration was adopted in 1998 and is "an expression of commitment by governments, employers' and workers' organizations to uphold basic human values – values that are vital to our social and economic lives."

The Declaration covers four areas:

• Freedom of association and the right to collective bargaining;
• The elimination of forced and compulsory labour;
• The abolition of child lobour, and;
• The elimination of discrimination in the workplace.

SA8000 (Social Accounting 8000), a private standard, created in 2000 by the Council on Economic Priorities Accreditation Agency (CEPAA), aims at improving the working conditions within companies and along the supply chain. It offers two possibilities to companies that want to demonstrate their commitment to social responsibility. Self-assessment is the first one and is meant to help companies, mainly retailers, implement a policy on social responsibility. The second possibility for companies is to certify their socially responsible employment practices according to the SA8000 standard. This second possibility, which is more stringent, is aimed at suppliers and manufacturers.

The SA8000 standard comprises nine key areas:

• child labour;
• forced labour;
• health and safety;
• freedom of association and collective bargaining;
• discrimination;
• disciplinary practices;
• working hours;
• compensation;
• management systems.

9.5.2   Covenants

Global Compact

The Global Compact is an international initiative launched by the General Secretary of the United Nations, Kofi Annan, on 26 July 2000. The Compact is composed of 10 principles in four categories and 'asks companies to embrace, support and enact, within their sphere of influence, a set of core values in the areas of human rights, labour standards, the environment and anti-corruption'. Companies commit themselves by signing the Global Compact (and implementing it!).

Human rights

- Principle 1: businesses should support and respect the protection of internationally proclaimed human rights.
- Principle 2: businesses should make sure that they are not complicit in human rights abuses.

Labour standards:

- Principle 3: businesses should uphold the freedom of association and the effective recognition of the right to collective bargaining.
- Principle 4: businesses should support the elimination of all forms of forced and compulsory labour.
- Principle 5: businesses should support the effective abolition of child labour.
- Principle 6: businesses should support the elimination of discrimination in respect of employment and occupation.

Environment:

- Principle 7: businesses should support a precautionary approach to environmental challenges;
- Principle 8: businesses should undertake initiatives to promote greater environmental responsibility.
- Principle 9: businesses should encourage the development and diffusion of environmentally friendly technologies.

Anti-corruption:

- Principle 10: businesses should work against corruption in all its forms, including extortion and bribery.

ETNO Sustainable Charter

The ETNO Sustainable Charter was launched in 2004 as an upgrade of the
ETNO Environmental Charter, signed by major European Telecom Network
Operators (ETNO) in 1996. It embraces the three pillars of the European Union
sustainable development strategy: environmental protection, social progress and
economic growth.

The charter signatories commit themselves to continuous improvement and
sharing best practices in the following areas:

- awareness;
- regulatory compliance;
- research and development;
- procurement;
- accountability;
- cooperation;
- management systems;
- employee relations.

Environmental Charter for the North American Telecommunications Industry
(the 'Environmental Charter')

This Charter was launched in 1999 by the United States Telecom Association (USTelecom), on the model of the European Charter and covers similar
topics. It was renamed the Environmental Sustainability Leadership Programme
Charter in 2005 and, based on the principles of the Environmental Charter,
serves to:

- encourage corporate commitment towards environmental sustainability;
- promote telecommunications products and services that contribute to environmental sustainability;
- demonstrate the industry's commitment to operating in an environmentally
  responsible manner;
- communicate industry progress on environmental sustainability issues;
- network with other national and international associations to share experiences and knowledge.

9.5.3   Tools

We present a few of the most common tools used within the framework of
corporate responsibility. These tools help to analyse impacts and outcomes or
provide guidance in reporting and communicating.

The following tools have been normalized or benefit from a large consensus:

- LCA – assessment of the impact of products during their life, from production to disposal.
- DfE – integration of environmental aspects in the design of products.
- GRI (Global Reporting Initiative) – an attempt to standardize non-financial reporting on a global scale, with a general reporting scheme applicable to any company. The third version (G3) was released in 2007.
- AA1000 (Accounting Assurance) – an attempt to normalize the external verification of reporting.

The following tools have not been fully normalized or are in development:

- *RA* (risk analysis) – risk analysis includes risk assessment and risk management. Risk assessment requires the identification of possible causes of damage, the probability of occurrence of damage and its magnitude should it occur.
- *Risk management* deals with measures and actions taken to either avoid the causes of risks (risk avoidance) or reduce the probability of occurrence or the magnitude of damage, or both (risks reduction). If the measures are not sufficient to eliminate the risks or cannot be applied, then the risks must be either accepted (risks retention) or insured (risk transfer), depending on the costs.
- The *ISO/IEC Guide 73* is a guide to risk management terminology and offers an overview of risk management practices. It was released in 2002.
- *TA* (Technology Assessment) is a procedure to comprehensively assess the complex social, ecological and economic implications of new technologies. Owing to the broad range of supposed impacts, TA most profitably combines an interdisciplinary approach and the advantages of different methods of investigation. A TA should at least combine an LCA to evaluate those influences that are certain and a risk analysis for potential, uncertain aspects.

These methods are applied to identify at an early stage technology-induced risks, the consequences of novel technologies for the society and the environment and their acceptance among the population. They can go as far as to assess the social, ethical or moral value of such novel technologies, depending on the scope assigned to the assessment or the criticality of the technology assessed (for instance those technologies dealing with human life).

Such assessments should deliver inputs and options for strategies, policies and decision-making help in order to balance and satisfy the interests at stake.

Not only the direct impacts are assessed, but also secondary or tertiary effects (see Section 9.6.2).

- *SM* (Stakeholder Management) refers to the identification of groups that have something common at stake with a company and the management of

activities to respond to the interests of these groups. It is a method of identifying differences in views and potential conflicts and of fostering common
understanding and resolution of issues through dialogue and involvement.
It aims at improving projects, products, services, processes and attitudes; it
should ultimately and ideally help companies to win the trust and confidence
of their stakeholders.

- *IM* (Issue Management) is a process to respond to and position companies
toward events that take place outside companies or toward stakeholder expectations that could affect their strategies, their core business or their reputation. Issue management and stakeholder management have similarities; they
are management tools of modern companies that demonstrate responsibility.

- *SC* (Score Card) is most effective when it is balanced and sustainable.
Scorecards are tools to translate vision into strategies and operational goals,
identify performance drivers and associate them with a set of measures and
targets as 'you can only improve what you can measure'. It is a 'dashboard'
for the conduct of a company. A sustainable scorecard is an extension of
the balanced scorecard, including further environmental and social areas
than the four classical areas that drive business performance (customers,
financial, operational, learning).

9.6 GUIDING CONCEPTS

The most important guiding concepts of sustainability were briefly addressed
in Section 9.2; they are listed below in more detail. They will help in designing
an effective management of corporate responsibility.

9.6.1 Triple Bottom Line

The triple bottom line, or TBL, refers to an enhanced accounting and reporting
concept that includes, besides traditional financial reporting, a reporting of the
social and environmental performance of a company. It is probably the major
guiding concept of corporate social responsibility and has even become its
embodiment (see Section 9.2).

Organizations, having recognized that sustainability is the most important
principle of action of responsible companies, must logically develop, maintain
and improve activities and processes aimed at reducing their negative environmental and social impacts. Consequently, they must engage in some form of
triple bottom line reporting, hence the purpose of the triple bottom line

The concept of a triple bottom line, although widely accepted under this
name and with this content, has been challenged as not being the most effective
method for putting companies onto the path to sustainability. It has been said

that, among other things, the analogy to established financial accounting is misleading or plain wrong, as concepts such as revenues and expenses, assets and liabilities, and gains and losses are yet to be defined for the accounting of social and environmental aspects [9] or simply do not make sense if these aspects are not monetarized.

Another issue raised is the fact that 'companies are pressured to think of CSR in generic ways instead of in the way most appropriate to each firm's strategy' [10]. This generic thinking drives companies to focus on irrelevant issues (to them) or on topics that they are not the most efficient in solving, or simply distracts investment from activities that would benefit both companies and society.

Let us rest this case for the moment by saying that responsible companies ought to identify both their negative and positive social and environmental impacts, integrate these aspects in their strategies and corporate culture and finally report their progress in mitigating or eliminating these negative impacts or claim credit for the improvement of positive impacts.

9.6.2   Levels of Effects

A typology of effects allows for a structured approach of impact analysis, better decision-making and meaningful responsibility. Here we follow the definition of the European Information Technology Observatory EITO (2002) on the classification of the impacts of ICT [11]. Impacts are classified into first, second and third orders.

First-order effects cover all direct impacts caused by ICT during the life of products, from manufacturing to disposal.

Second-order effects are the consequences of the application of ICT, resulting in higher process efficiency when used in other industries or business systems, higher eco-efficiency (MIPS), energy efficiency or even dematerialization.

Third-order effects result from changes in behaviour and demand patterns of large numbers of people using ICT. These changes could generate a rebound effect.

The magnitude of indirect effects appears to be of medium to large importance, whereas the magnitude of direct, first-order effects seems comparatively small. Under this perspective, acting on the causes of indirect effects will bring the largest benefits in curbing the impacts over time.

A detailed list of first, second and third order impacts is given in Section 9.8.

9.6.3   Equity

Equity is the fundamental social component that, together with economic growth and environmental maintenance, defines sustainable development. Equity derives from social justice and can be understood in the context of sustainable development at least as a minimum level of income and environmental quality that should be available to everyone.

Sustained equity will be realized in the present and maintained in the future, to 'satisfy today's needs without limiting the satisfaction of future generations' needs'. One speaks then of intragenerational and intergenerational equity, the latter being probably the least explained issue of sustainability, and one that makes this concept difficult to apply.

9.6.4   Time

Time is an essential element of sustainability, its 'fourth' dimension. Sustainable development requires prediction of the future, which by definition is uncertain (only hindsight is an exact science!). As forecasting is burdened with uncertainties, it may help to select activities proven to be sustainable and discard those that are unsustainable as a possible way to reduce uncertainties and prepare for a better future.

The time factor intervenes in intergenerational equity (the definition of sustainable development given by the Brundtland commission) and in the implication that benefits should last, to fit with the concept of sustainability. Consequently, this will force us to live on the interest only and not on the capital or, at the very minimum, the rate of capital exhaustion should be extremely small (not quite realized in the present situation!).

9.6.5   Efficiency

One fundamental guiding concept of sustainability refers to increases in efficiency, a first step in significantly improving resource productivity and promoting resource-efficient technologies to help reduce our ecological footprint.

Efficiency is an indicator of resource productivity or performance expressed as a ratio of two terms, a term of 'value creation' over a term of resources used. These two terms could also be taken in a reverse manner as an 'intensity' ratio. A trivial example is the performance description of car engines, given in miles per gallon in the United States (an efficiency – the more the better) or in litres per (100) kilometre(s) (an intensity – the lower the better) in Europe or the rest of the world.

Efficiency, known by different names, all referring to the same basic concept, was intensively developed in the 1990s and is enjoying a revival as this book goes to press. Eco-efficiency is a term coined and defined in an extremely long sentence by the World Council on Sustainable Development (WBCSD) in 1992. Concisely, eco-efficiency is the creation of more value while causing less impact [12]. Eco-efficiency was made popular as a way to 'double wealth while halving the resource use' [13]. This increase in efficiency by a factor of 4 was proposed in a well-known book in the late 1990s. On the very same model, a more ambitious factor of 10 was suggested as the minimum necessary improvement in order to approach ecological sustainability. Finally yet importantly, linked with the factor 10 concept, a tool measuring the material intensity per unit of service (MIPS) appeared at the same time [14].

A common measure of the efficiency of companies is the ratio of 'turnover per total energy demand' in currency unit per Joule, sometimes presented as an intensity in Joules per currency unit. ISP and network operators can control and compare their efficiency by measuring their 'traffic efficiency', defined by the number of bits transmitted per energy demand (bits/J) or, when focusing on server farms and IT-centres, their 'capacity efficiency' in bytes per energy demand (bytes/J).

Notice that these eco-efficiencies are restricted to one aspect of efficiency only, namely the use of energy to generate value. Other impacts are not taken into consideration. This is perhaps acceptable in a first approximation of eco-efficiency as the energy demand by the ICT sector represents its main impact and therefore could be taken as a proxy for the resources used in the creation of 'value'.

9.6.6   Limits and Carrying Capacity

The triple bottom line concept and the typology of impacts provide the scope of corporate responsibility. Nevertheless, it is necessary to consider a set of targets or limits when designing sound sustainability strategic paths. These limits derive from the carrying capacity, defined as the maximum population density (human, fauna and flora) which can be supported in a given area within the limits of its natural resources and without degrading its environment for future generations of this population.

The productive surfaces on Earth total about 11.4 billion hectares for a world population of 6 billion humans. This represents 1.9 ha per capita, about two football fields per person, the actual carrying capacity. Globally, from the model presented in Section 9.1, the actual ecological footprint is about 2.3 ha per person (with very large geographical disparities), 17 % above the possible carrying capacity. This is a very unsustainable situation in the long run.

Corporate responsibility is demonstrated, for instance, when organizations, integrating the concepts of the carrying capacity and the ecological footprint in their sustainability strategies, select as far as possible their energy systems accordingly: the ecological footprint of liquid fossil fuel of 1.4 ha per 100 GJ must be taken into consideration and balanced against the ecological footprint of, say, wind energy (0.008 ha/100 GJ) [15].

9.7   CORPORATE RESPONSIBILITY AND NGN

It is envisioned that the NGN ahead of us will bring new levels of comfort and life experience. In order to fulfil these goals and to completely exploit the potentials of NGN, it is necessary to clarify the new challenges faced and their possible impacts: in the end, the negative aspects will not undermine the promised comfort and poison the expected thrilling life experiences. The positive aspects will largely exceed the negative ones or these will disappear: under the bottom line we must find only 'positive sums'.

As highlighted in Chapter 1, the ICT industry finds itself in a situation where offer surpasses demand (see Figure 1.1). Three possibilities exist: simply push the demand, reduce the offer or tailor the technology to the exact needs of customers (the key proposal of the authors of this book). A partial apprehension of the situation, economic, environmental or social only, will clearly have flaws; only a comprehensive, sustainable approach has merits.

In order to choose the proper behaviour that is compatible with sustainability, one needs to identify the benefits and the impacts of NGN. We have argued in Chapter 1 that the ICT world must now apply a model centred on the customers' expectations to fulfil their needs. The 'right' behaviour (knowing the needs of customers, knowing the impacts and benefits and knowing what to do) will be the key to building up trust, acceptance and use of NGN by customers and stakeholders.

Applying this approach, that is, identifying and satisfying the needs of customers, is also a way to reduce risks and prevent the always possible rejection of NGN. Technology should help its users, not subject them: abiding by this ideal could make risk analysis and technology assessment unnecessary.

9.7.1   Balancing the Benefits and Impacts of NGN

Next generation networks are networks that will link customers, terminals, sensors, machines and products of all kinds. NGN are characterized by their bandwidth adaptability, real-time operations, access technology agnosticism, pervasiveness (service access everywhere) and ubiquity (networking) offered to users (human or machines). In short, generation, flow, computation and storage of data will be at the heart of NGN and its key ingredients.

An as yet unseen combination of bandwidth, computing power and storage capacity will permit improvements in the quality of life of people and the efficiency of companies. Individuals, communities and companies, thanks to the open access and the standardized service enablers, will have the power and the possibilities to develop their own services or services that can be offered to everyone.

The ICT industry has already taken the first steps in this direction. Typical NGN tools such as web 2.0 and IPv6 are being introduced or will be introduced shortly; the age of access is approaching as most barriers are disappearing and behaviour patterns are evolving; the eco-efficiency of operators, measured by any ratio of traffic or capacity per unit of energy demand, has increased manifold over the past decades, slowly before and tremendously after the opening of the telecommunications market in the mid-1990 and the mass merging of IT and CT at the same time.

These trends are gaining momentum and magnitude. One can witness an explosion of new projects and the development of established fields of activities, which are characteristics of the coming decade. Intelligent homes can be visited in many university laboratories; portability and access agnosticism has realized; terminals are readily available, cheap and reliable.

Related to this latter aspect, terminals in the form of machines and robots were identified by some as the next big technology revolution. Science fiction in the 1950s, a reality in industry since the mid 1970s, robots began to pervade every aspect of life one generation later. Thanks to available capital and by providing assistance in the form of specific programming tools and software (a robotic programming kit is already commercialized), a famous company is investing in this relatively new field, trying to re-edit its exploit of controlling an entire industry, this time by linking the machines to and directing the flow of data toward the PC found in every home [16].

It is both entertaining and interesting to see the different perspectives on NGN: while network operators focus on the network technologies, the IT world has a tendency to speak mainly of pervasive computing whereas enlightened entrepreneurs are already starting to exploit almost unexplored territories. This confrontation of points of view can prove very fruitful for responsible companies.

In the following paragraphs, we underline three issues with a potential for conflicts of interests, centralization, complexity and acceptance.

Centralization

Concerning terminals, the debate is particularly interesting. Should one centralize or distribute the 'intelligence'; should it be in the network (servers) or in the terminals? This, of course, has consequences in terms of handling and protecting data and the complexity of management and responsibility, but, as terminals could be more or less complex, also in terms of energy and resources consumption or even in terms of social impact (autonomy vs dependency).

Adopting thin client technology presents the advantage of disseminating simpler, less energy- and material-intensive terminals, and reducing manufacturing complexity and electronic waste. It offers the possibility to share central, powerful, up-to-date computing resources. The downside of this solution is primarily the impact of the energy consumption of server farms. This could be curbed in two ways, through the design of chips or through the better use of servers, today mostly dedicated to only one application or by choosing renew-able energies over fossil ones. Concentrating 'intelligence' and data repositories may also raise security issues as they could become targets for malevolent people.

Complexity

Offering more choice and possibilities, tailoring them to the individual, multi-plying services and computing pervasiveness are desirable to improve quality of life. Uncontrolled or purposeless, products, services and processes could rapidly become too difficult to operate for the average customers and create unneces-sary barriers. Services such as the Web should provide paths for the diffusion of knowledge and information and not entangle individuals into mazes, restricting their freedom. Finally, distributed ICT systems may evolve toward structures that are more complex than previously intended, adding a loss of overview to the management challenge.

It has been observed that large systems with a large number of components have a tendency to evolve towards a critical intermediate state, which could be far from the equilibrium. A minor perturbation, without the help of external agents, may have catastrophic consequences and result in a new organization (self-organized criticity) that is not foreseen and out of control.

Assigning responsibility within complex systems may prove almost impos-sible, due to not only their possible evolution, but also resulting from a certain dilution of decision centres typical of distributed systems. The combination of web 2.0 and IPv6 will at last authorize ubiquitous peer-to-peer communica-tion. Machines will command other machines, directly and without centralized control: which responsibility will a machine or a robot have? More than 50 years ago, the 'three laws of robotics' were proposed in entertaining novels by Isaac Asimov as a type of 'responsibility' – before its time and not totally out of date today!

Acceptance

Offer surpasses demand. If the needs of customers are not or too weakly taken into account, new services and products will be ignored and the industry will not recoup its investments. A model driven by the customers' needs is key to solving this dilemma (see Chapter 1). Ignoring the imperative to consider the customers' needs, it will be tempting to fall back into the usual trap of 'offer creates demand' and meaninglessly push the demand. Measuring the publicity budget and spending of corporations can quite easily detect this situation. Another, more subtle, way to push demand is to control the demography of products, by bringing them rapidly out of fashion and imposing their replacement instead of repairing them. This attitude may generate cash, but surely also a lot of waste.

These topics of centralization vs distribution, complexity vs simplicity and finally acceptance vs reluctance combine the new mix of issues that proponents of ICT will face. Although not typical of ICT, it will nevertheless be very acute in this economic sector by reason of the magnitude of the issues involved.

9.7.2   The Positive Aspects

The tools presented in Section 9.5, especially risk analysis, life cycle analysis and technology assessment, are adequate for performing an overall analysis. Within the framework of a corporate culture and values, the application of the principles of precaution and responsibility will offer the necessary guideline to deal with the issues identified.

Let us start with the positive aspects brought by ICT that NGN will amplify. The undisputed successes of ICT lie in four key categories:

* increase of efficiency;
* dematerialization;
* access to information and knowledge;
* traffic substitution

Efficiency

The increase in efficiency is perhaps one of the main achievements of the ICT sector. Eco-efficiency, measuring traffic per energy demand in bits per Joule, has doubled over the past decade, from about 100 bits/J in 1991 [13] to more than 300 bits/J in 2006, a factor of 3 in only 15 years.

Telecommunications is less energy-intensive and CO_2 emitting than many other industries. It is also efficient in comparison: as an example, telecom generates 2.29 % of UK economic output, with only 0.82 % of the sector's total costs coming from energy [17].

ICT have a major impact from their use in other business. Efficiencies in industry, in building, in production and in planning can be traced back to an increased usage of ICT. ICT bring a time compression that, coupled with cultural changes (accepting and using new technologies at work), drives productivity up. The strategic use of ICT can further contribute to efficiency: a clever combination of a company's databases combined with artificial intelligence can help and support quick and nevertheless robust decisions and reduce uncertainties (in the hands of experienced and critical management).

The PC industry has tremendously improved the computing efficiency per unit cost over the past 40 years, nicely following the observation of Gordon Moore. On the other hand, considering the energy efficiency of computers and servers, the result is less rosy as they waste today between half of their energy (for PCs and one-third for servers), pushing IT operators to force-cool their servers farms, which is not a very productive use of valuable energy. Nevertheless, new technologies (thin clients for instance), better designs and multiparameter optimization combined together bring the energy consumption down to about 15 W per unit, including the server share, compared with a typical 150–200 W per PC.

Dematerialization

Dematerialization in term of MIPS is quite noticeable. Whereas 80 kg of copper per line and per kilometre were necessary in 1915 to carry a signal, only 0.01 g of glass are sufficient today, a factor of 80 million.

Mobile phones, including their batteries, are down from 5 kg in 1984 to about 50 g in 2006, a continuous improvement representing over time downsizing by a factor of 100. Television, video and computer displays have followed a different pattern of dematerialization and energy efficiency with the abandonment of cathodic ray tubes (CRT) almost overnight and their replacement with flat panel displays.

The signals previously transported on wires are now transmitted wirelessly over the air. This change in medium represents another type of 'dematerialization' that occurred over the past 20 years for telecommunications.

Access

Under access to information and knowledge, besides the obvious usage of the Internet, we include monitoring, reporting, alerting and responsive services, for security purposes or for specific uses in hostile environments, both natural and social.

Traffic Substitution

Traffic can be replaced by ICT. Conferencing services offer today a high level of comfort; the virtual spaces of tomorrow will make business travel unnecessary. Tele-work and other teleservices already contribute to a non-negligible reduction of CO_2-emissions, allowing people to work in a flexible way, experiencing a richer family and social life in parallel.

A joint initiative by the association of European Telecoms Network Operators and the World Wildlife Fund resulted in the publication of a report named *Saving the climate @ the speed of light: ICT for CO_2 reductions* on the potentials, opportunities and strategies for fighting against climate change using ICT [18].

The ETNO/WWF report outlines a roadmap for the ICT sector, and sets out a target for 2010 to use ICT to reduce CO_2 emissions by 50 million tonnes. Potentials savings per service were estimated:

- video conferencing – if 20 % of business travel in the EU 25 was replaced by video conferencing, this would save 22.3 million tonnes of CO_2;
- audio conferencing – if 50 % of EU workers replaced one meeting with one audio conference a year, this would save 2.2 million tonnes of CO_2;
- flexi-work – if 10 % of the EU 25 workforce were to become flexi-workers, this could save 22.17 million tonnes of CO_2 per year;
- on-line billing – 100 million customers receiving on-line phone bills would save 109,100 tonnes of CO_2;
- Web-based tax returns – 193 million web-based tax returns would save 195,000 tonnes of CO_2.

9.7.3 The Challenges Ahead

Some challenges must be overcome to ensure the success of NGN and its acceptance in order to fulfil its promise of a better life and richer customer experiences. The barriers to be lowered are many, e.g. educational, technical and financial. We address some points typical of the ICT sector:

- availability;
- energy demand;
- rebound effect;
- electronic waste (e-waste);
- NIR exposure (non-ionizing radiation exposure);
- data protection;
- pervasiveness.

Availability

A trivial challenge is of technical and economic nature: the key to successful NGN resides in the standardization of interfaces, the availability of computing power and the ubiquity of such elements. Differential introduction of NGN may exacerbate the digital divide, slow down its acceptance and not bring fast enough the social as well as the financial benefits necessary to sustain its development.

Energy Demand

The underlying limiting factor is of a physical nature: there may not be enough energy to power the networks and terminals at the present rate of energy consumption. The ICT devices will be always connected, most of the time in an idle state (standby mode, sleep mode or hibernation), and nevertheless consume power unproductively. To remain credible, the ICT will have to bring efficiency to other economic sectors not at the cost of its own efficiency. Moreover, with the energy production still fossil-fuel-dependant, the ICT indirectly emits quite a lot of CO_2, whereas its services bear the greatest potential to reduce CO_2 emissions from traffic. The ICT industry may well shoot itself in both feet if it does not swiftly tackle its energy and emissions problems.

Responsible companies have identified this issue and have started to solve the contradiction between not so clean operations and commercial offers of 'green' and efficient services. Reduction of energy consumption, sale of excess heat to neighbours, purchase of or own production of green electricity (photovoltaic, wind, water or geothermic) are measures commonly taken, even going further to compensate for CO_2 emissions by putting to work the flexibility mechanisms of the Kyoto protocol (a lot of ET, a little bit of CDM and almost no JI).

Rebound Effect

Efficiency increases have a pernicious consequence: it has been observed that more demand is generated, reducing the previous gain. This 'rebound effect' can be explained by economic theory (it is related to the price elasticity of the resource at stake). As efficiency gains in the ICT are rapid (see Moore's law), the rebound effect follows a similar pace.

Electronic Waste

The high churn rate due to efficiency gains and the rebound effect push up the consumption of electronic devices and increase electronic waste. The amount of e-waste is alarming: mobile phones are replaced every second year or faster. PCs are kept for 3–5 years and small entertainment devices are extremely short-lived. Fashion makes many victims nowadays and quality equipment goods have become mere consumption goods, wasting extremely valuable technologies.

The ICT sector will soon find advantages to introducing a closed-loop economy by recycling more of its products, as some key materials are not abundant or are linked with very negative social aspects (notably tantalum, also known under the name Coltan).

A political decision of the EU, concretized by a directive banning six hazardous materials [lead, mercury, cadmium, hexavalent chromium and two polybrominated fire retardants, in the Directive on the Restriction of Hazardous Substances Directive (RoHS) 2002/95/EC] from electrical and electronic equipment has had an impact on the manufacture and disposal of such products. It is part of a solution to the problem of e-waste.

NIR

Exposure to non-ionizing radiations is likely to increase with the NGN. This issue still raises fear among some people and could have the potential for conflicts. Thermal effects are well known, although they appear to be below the limits set by most legislations. Other possible effects from short-term exposure have been thoroughly investigated and ruled out (no effects or no replication of anecdotic effects). Long-term exposures have been less investigated and no conclusion can be drawn at the moment. Nevertheless, it remains a sensitive topic and should be handled with care and empathy.

Data Protection

Huge amounts of data will be created, handled, transmitted and stored. This information will represent a tremendous value that will require a high level of protection, and is perhaps redundant (this is another cause of energy consumption that is never addressed).

Who is conscious of the quantity of his or her personal data available on the net? Who can access this data, who manages it, who has responsibility over it? There are ectoplasms 'in the machine' and everyone, without always realizing it, has a double life! Access, consumption habits, transactions, etc. are possibly scrutinized, as the value of data is not only directly financial but resides in pattern analysis and the resultant predictability of behaviour.

Pervasiveness

The ubiquitous presence of active devices could develop fear, distrust and an uneasy feeling of being constantly under surveillance among users and non-users. Rejection of NGN will result if privacy or guarantees of personal freedom are missing.

9.8 SUMMARY OF IMPACTS

The effects of ICT have been structured in a conceptual framework with three levels of impact (see Table 9.1 and also Section 9.6.2). First-order-level impacts are direct impacts from the activities of a company. Responsibility over this level of impacts is in principle clear, taken into account by the relevant legislation and today internalized by companies. The second and third levels are indirect consequences of the activities, products and services of ICT companies. They cannot be directly accountable for those impacts, but may be involved as the magnitude of these impacts is much larger than that of the first-order level.

Table 9.1 Order levels of impacts from the use and applications of ICT [11]

	First-order level	Second-order level	Third-order level
Environmental impacts	Impacts from energy use such as GHG emissions and land use Other resource use impacts such as water consumption, wastewater discharge, air emissions, solid waste discharge Biodiversity impacts Toxic release impacts	Environmental impacts of increased or decreased transport Environmental impacts from increased product life-times Impacts from the use of ICT in other business systems Environmental impacts of the rebound effects created by a change in consumer behaviour	Energy use impacts such as carbon emissions stemming from aggregated ICT use at the macro economic level Possible environmental impacts on regions due to changing settlement patterns Land-use impacts from aggregated ICT use at the macro economic level
Social impacts	Increase in employment Occupational and customer health and safety impacts Intellectual property rights impacts Customer privacy impacts (can be also a second-order effect)	Social impacts from access to better services/digital divide (health services, e-learning, telework) Social impacts/risks from security challenges Social impacts from addressing barriers to access Social impacts from enhancement of democratic participation Availability in emergency situations and disaster relief Impacts from having new communities on-line	Cultural homogeneity Cultural biodiversity Enhancement of local communities Enhancement of civic culture
Economic impacts	Net sales Payroll and personnel benefits Liquidity impacts (such as debt, borrowings, dividends position) Subsidies received Tax exemptions or decrease in tax Impacts on community, civil society and other groups in terms of donations Impacts on other geographic locations and low-income groups	Economic impacts from application of ICT within other businesses Enhancement of innovation and competitiveness Establishment of new financial markets enabling growth and wider participation Increase in empowerment of consumers Encouragement of ethical corporate behaviour	Economic impacts from long-term and fundamental changes to the global economy Economic impact on patterns of wealth Better share of economic wealth

9.9   IN A NUTSHELL

In conclusion, clever, responsible companies have not only recognized the complexity of the global environment they now face everyday; they have also acknowledged that customers, stakeholders and shareholders alike expect them to undertake the necessary steps to develop their business in a sustainable manner. This in turn requires considering the social, environmental as well as the economic consequences of activities, listening to stakeholders and assessing new technologies for risks and opportunities. This attitude reflects the new paradigm of sustainable development.

Norms and tools are available to guide companies towards more responsible behaviour. Blindly following them may bring some benefits. It helps far more to recognize that priorities and solutions are dictated by the situation and context surrounding companies.

Responsible companies will then minimize their negative impacts. Truly responsible companies will adopt a strategic attitude and identify their positive impacts to amplify them, for the greatest benefits of both their customers and the society.

Energy consumption, electronic waste, lack of privacy and dependency are among the dangers challenging the opportunities NGN will bring in access to information and education, efficiency, dematerialization and services with the potential to curb climate change.

REFERENCES

[1] WWF. *The living Report*, 2006; 51.
[2] Gigi Richard. Human Carrying Capacity of Earth. *ILEA Leaf*, Winter 2002 issue.
[3] The present reduction of biodiversity is now officially considered the sixth mass extinction event, in reference to the five major crises of extinction of species experienced by the Earth. Source: *Report on Global Biodiversity, 2006*. UN Conference on Biodiversity, 2006. 'Human activities are putting such strain on the natural functions of earth that the ability of the planet's ecosystems to sustain future generations can no longer be taken for granted', *UN Millennium Assessment*, 2005.
[4] World Commission on Environment and Development (WCED). *Our Common Future*. Oxford University Press, 1987
[5] John Elkington. *Cannibals with Forks: the Triple Bottom Line of 21st Century Business*. New Society Publishers, 1998.
[6] Figures used: freshwater estimate equals 2.5 % of world's water, about 35 million km^3; source: UNEP vital water graphics. Population estimate 2006: ~6 billion; 1876, ~1.3 billion. Source: US Census Bureau.
[7] Joëlle Stolz. La 'crise de la tortilla' fait descendre les Mexicains dans la rue, *Le Monde*, 2 February 2007.
[8] Gary Hamel, C.K. Prahalad. *Competing for the Future*. HBS Press, 1994.

[9] Wayne Norman, Chris MacDonald. Getting to the Bottom of 'Triple Bottom Line', *Business Ethics Quarterly*, April 2004.

[10] Michael Porter, Mark Kramer. Strategy and Society, the link between competitive advantage and corporate social responsibility. *Harvard Business Review*, December 2006. The article, although focused on the social aspects of corporate responsibility, offers a method that can be applied to the environmental aspects as well. A mapping of the social impacts of the value chain and the identification of the social influences on competitiveness is provided.

[11] Diverse sources were used for the typology and classification of impacts of ICT; see: European Information Technology Observatory (EITO). *The Impact of ICT on Sustainable Development*, 2002; 253. Andreas Köhler, Lorenz Erdmann. Expected Environmental Impacts of Pervasive Computing. *Human and Ecological Risk Assessment* 10: 831–852, 2004. Michael Kuhndt, Burcu Tunçer, Christa Liedtke. A triple bottom line innovation audit tool for ICT product-service mix applications. *2nd International Workshop on Sustainable Consumption*, Tokyo, 12–13 December 2003.

[12] World Business Council on Sustainable Development (WBCSD). *Eco-efficiency, Creating more Value with Less Impact*, Geneva, October 2000.

[13] Ernst Ulrich von Weizsäcker, Amory and Hunter Lovins. *Factor Four – Doubling Wealth, Halving Resource Use*. Earthscan: London, 1998.

[14] F. Schmidt-Bleek. *Wieviel Umwelt braucht der Mensch – MIPS, das Maß für ökologisches Wirtschaften*. Birkhäuser: Basel, 1993.

[15] Mathis Wackernagel, William Rees. *Our Ecological Footprint, Reducing Human Impact on the Earth*. New Society Publishers: Gabriola Island, British Columbia, 1996.

[16] Bill Gates. A robot in every home. *Scientific American*, 296, January 2007; 44.

[17] Nicholas Stern. *The Economics of Climate Change, the Stern Review*. Cambridge University Press, 2006; 297, Annex Table 11 A.1.

[18] WWF. *Saving the Climate @ the Speed of Light: ICT for CO_2 Reductions. First Roadmap for Reduced CO_2 Emissions in the EU and Beyond*. October 2006.

10

Summary

NGN is introduced in this book as an answer to the life-and-death challenges faced by traditional telecom operators today. The answer opens new prospects; these are deepened with a long-term view of NGN that will enable the coming of another revolution in the history of human beings in improvement of quality of life and new life experiences, comparable to the industrial revolution.

The NGN's time has arrived; the pioneer work of ITU has marked this start. Even though many aspects are far from being crystal clear, the road ahead can already be discerned:

- Traditional telecom operators will beat the challenges with a smart designed network evolution that leads to a customer need-driven operation approach, optimized for its own conditions and purposes.
- Cable network operators will extend television and radio businesses by exploring new business opportunities within the existing broadband cable connection to home.
- Internet provider will go beyond their 'best-effort' image to become carrier-grade providers by stepping into the CT business with new wireless networking technologies or by cooperating with the existing network operators.
- New players will see the tremendous opportunities coming with NGN in the domain of transport, services, applications, content and information.
- Researchers will grasp the fields needing to be explored for NGN development.

Next Generation Networks: Perspectives and Potentials   Jingming Li Salina and Pascal Salina
© 2007 John Wiley & Sons, Ltd

NGN is based on ICT with high-level technology and management, providing:

- *Ubiquitous connectivity* with the co-existing access technologies from narrowband to broadband and from supporting portability to nomadic to high mobility; including wired, fixed wireless, mobile and satellite; covering space, terrestrial and underground, on-the-water and underwater, and everywhere.
- *Pervasive accessibility* to services, applications, content and information with any device remotely configurable and software-upgradeable.
- *Prosperous IP-based e-applications* substituting today's telecommunications and broadcasting services with smart features, satisfying the needs of humans from every corner, far beyond what we can imagine today and deserving to be called the electronic and software revolution.
- *Reciprocity of service consumers and providers* to allow every customer to provide and share her/his applications, content and information with others, applying sophisticated communication technology.
- Face-to-face human-like distant communication using the five senses and context information without language barriers.
- Virtual living environments joined to physically separated life spaces.
- Human-like user interfaces for service activation and semantic searching.
- Biometric-enabled access authentication.

NGN will have real-time interactive control and management functions at both service and transport layers:

- to enable resource-assignment for the network taking into account the business importance of a service, a customer, the QoS requirement, etc.;
- to enable proactive service and network management to minimize the operational problem level;
- to enable managed QoS granularity and security levels to provide the adequate QoS and security to the end-user with service delivery;
- to enable SLA-based customer management, where a customer pays for the value he or she gets;
- to enable the flexibility to add, remove or maintain

  o services, applications, content and information from third-party providers;
  o access transport networks based on the same or different technologies, from the same or different operators.
  o core transport networks based on the same or different technologies, from the same or different operators.

NGN will enable a richer and extended landscape of providers, operators and customers. It will bring tremendous advantages to companies and individuals, in terms of access to information, education and knowledge, efficiency and dematerialization. Communities, social networks and changes in behaviour patterns will enrich the life of many; remote and virtual worlds will be accessible, offering new experiences.

Attractive as this wealth of prospects may be, the side effects that the ICT will bring along their development should not be underestimated. Energy consumption, electronic waste, digital exclusion and the fear of electromagnetic fields are important issues to tackle and to solve as early as possible, to gain and maintain credibility and confidence among customers, stakeholders and shareholders.

NGN will remain as one of the busiest fields for many years to come, as it will have a major impact on our lives. The final NGN will be born in the noise of discussion.

Last but not least, we stress again the NGN Perspectives and Potentials:

- *Next generation networks will network persons, devices and resources independently of distance, location and time, through integrated intelligent interfaces and with enriched media.*
- *Next generation networks as a platform will offer ubiquitous connectivity and intelligent interfaces for human and machine communication as well as pervasive service access, bringing value to human life for its improvement and new experiences. This platform will also provide a 'playground' for everybody to create and deliver services to others.*

Glossary

2D	Two-Dimensional
2G	Second Generation
3D	Three-Dimensional
3G	Third Generation
3GPP	Third Generation Partnership Project
3GPP2	Third Generation Partnership Project 2
4G	Fourth Generation
AAA	Algorithm Architecture Adequation
AA1000	Accounting Assurance 1000
ADSL	Asymmetric Digital Subscriber Line
AF	Assured Forwarding
A-GPS	Assisted GPS
aGW	access Gateway
AIPN	All IP Network
ANI	Application Network Interface
API	Application Programming Interface
ARF	Access Relay Function
ARIB	Association of Radio Industries and Businesses (Japan)
ASF	Application Server Function
ATIS	Alliance for Telecommunications Industry Solutions
ATM	Asynchronous Transfer Mode
BAN	Body Area Network
BGF	Border Gateway Function
BGP	Border Gateway Protocol
BPON	Broadband Passive Optical Networks
BRAN	Broadband Radio Access Networks
BSS	Business Supporting System
CAMEL	Customized Application for Mobile Network Enhanced Logic
CAN	Car Area Network
CCSA	China Communications Standard Association (China)
CDM	Clean Development Mechanism (Kyoto Protocol)
CDMA	Code Division Multiple Access

CEPAA	Council on Economic Priorities Accreditation Agency
CI	Contract Interface
CJK	China Japan Korea
CN	Core Network
CTE	Customer Terminal Equipment
CRM	Customer Relationship Management
CRT	Cathodic Ray Tubes
CSR	Corporate Social Responsibility
CT	Communication Technology
DAB	Digital Audio Broadcasting
DAN	Device Area Network
DECT	Digital Enhanced Cordless Telecommunications (former for Digital European Cordless Telephony)
DfE	Design for Environment
DiffServ	Differentiated Services
DMB	Digital Multimedia Broadcasting
DMB-T	Terrestrial Digital Multimedia Broadcasting
DMTF	Distributed Management Task Force
DRM	Digital Rights Management
DSCP	DiffServ Code Point
DVB	Digital Video Broadcasting
DVB-C	DVB Cable
DVB-H	Handheld Digital Video Broadcasting
DVB-HS	DVB Handheld and Satellite
DVB-S	DVB Satellite
DVB-T	Terrestrial Digital Video Broadcasting
DVD	Digital Video Disk
EDGE	Enhanced Data Rates for Global/GSM Evolution
EF	Expedited Forwarding
EGPRS	Enhancd GPRS
EITO	European Information Technology Observatory
EMAS	Eco-Management and Audit Scheme
EMS	Enhanced Messaging Service
eNB	evolved Node B
EPC	Evolved Packet Core
EPE	Environmental Performance Evaluation
ePDG	enhanced Packet Data Gateway
EPON	Ethernet Passive Optical Networks
eRAN	evolved Radio Access Network
ET	Emissions Trading (Kyoto Protocol, CO_2 and other Gaseous Pollutants)
ETNO	European Telecommunication Networks Operation Association
eTOM	enhanced TOM
ETSI	European Telecommunications Standards Institute
EU	European Union
eUTRAN	evolved Universal Terrestrial Radio Access Network
EVDO	Evolution Data Optimized

FE	Functional Entity
FFS	For Further Study
FG-NGN	Focus Group-NGN
FMC	Fixed Mobile Convergence
FRAND	Fair, Reasonable and Non-Discriminatory
FTP	File Transfer Protocol
FTTH	Fibre To The Home
FTTx	Fiber To The x (x = home, building, curb)
FW	Framework
GEO	Geostationary Earth Orbit
GERAN	GSM/EDGE Radio Access Network
GGSN	Gateway GPRS Support Node
GMPLS	Generalized MPLS
GOCAP	Generic Overload Control Activation Protocol
GPON	Gigabit Passive Optical Networks
GPRS	General Packet Radio Service
GPS	Global Positioning System
GRI	Global Reporting Initiative
GSI-NGN	Global Standards Initiative-NGN
GSM	Global System for Mobile Communication
GTP	GPRS Tunnelling Protocol
GUP	Generic User Profile
GW	Gateway
HAN	Home Area Network
HDTV	High-Definition Television
HEF	Human Ecological Footprint
HSDPA	High-Speed Downlink Packet Access
HSPA	High-Speed Packet Access
HSS	Home Subscriber Server
HSUPA	High-Speed Uplink Packet Access
IBCF	Interconnection Border Control Function
ICT	Information and Communication Technology
ID	Identification
IEEE	Institute of Electrical and Electronics Engineers
IETF	Internet Engineering Task Force
IM	Issue Management
IMS	IP Multimedia Subsystem
IN	Intelligent Network
InRD	Infrared
IntServ	Integrated Services
IP	Internet Protocol
IP CN	IP Core Network
IP-CAN	IP-Connectivity Access Network
IPR	Intellectual Property Rights
IPsec	secured IP
IPTV	Internet Protocol Television
IPv4	IP version 4

IPv6	IP version 6
ISDB-T	Integrated Services Digital Broadcasting-Terrestrial
ISDN	Integrated Services Digital Network
ISO	International Standardization Organization
ISP	Internet Service Provider
ISUP	ISDN User Part
IT	Information Technology
ITIL	IT Information Library
ITU	International Telecommunication Union
IUCN	Internationa Union for the Conservation of Nature
IWF	Interworking Function
JI	Joint Implementation (Kyoto Protocol)
JRG-NGN	Joint Rapporteur Group on NGN
KPI	Key Performance Indicator
KQI	Key Quality Indicator
L2TF	Layer 2 Termination Function
LAN	Local Area Network
LCA	Life Cycle Analysis
LEO	Low Earth Orbit
LMDS	Local Multipoint Distribution System
LSDI	Large Screen Digital Imagery
LSP	Label Switched Path
LTE	Long-Term Evolution
MAC	Media Access Control
MBMS	Multiple Broadcast Multimedia Services
MediaFLO	Media Forward Link Only
MEO	Medium Earth Orbit
MGF	Media Gateway Function
MIMO	Multiple Input Multiple Output
MIPS	Material Intensity Per unit of Service
MIPv4	Mobile IP version 4
MIPv6	Mobile IP version 6
MMDS	Multichannel Multipoint Distribution Systems
MME	Mobility Management Entity
MMoIP	Multimedia over IP
MMS	Multimedia Message Service
MOS	Mean Opinion Score
MoU	Memorandum of Understanding
MPLS	Multi-Protocol Label Switching
MPLS-T	MPLS Transport
MPtMP	Multipoint to Multipoint
MRFP	Media Resource Function Processor
MVDS	Multichannel Video Distribution System
NACF	Network Attachment Control Functions
NAPT	Network Address Port Translation
NASS	Network Attachment Subsysem
NAT	Network Address Translation

NE	Network Element
NFC	Near-Field Communication
NGMN	Next Generation Mobile Network
NGN	Next Generation Networks
NGN-GSI	NGN Global Standards Initiative
NGN-MFG	NGN Management Focus Group
NGOSS	New Generation Operation System and Software
NIR	Non-Ionizing Radiation
NNI	Network Node Interface
OAM	Operation And Maintenance
OCAF-FG	Open Communication Architecture Forum–Focus Group
OFDM	Orthogonal Frequency Division Multiplex
OFDMA	Orthogonal Frequency Division Multiple Access
OMA	Open Mobile Aliance
OS	Operating System
OSA	Open Service Access
OSS	Operation Supporting System
OTA	Over The Air
PAM	Presence and Availability Management
PAN	Personal Area Network
PBX	Private Branch Exchange
PC	Personal Computer
PCEP	Policy and Charging Enforcement Point
PCRF	Policy and Charging Rules Function
PDA	Personal Digital Assistant
PES	PSTN/ISDN Emulation Subsystem
PLC	Power Line Communication
PLMN	Public Land Mobile Network
PON	Passive Optical Network
POTS	Plane Old Telephony Service
PSTN	Public Switched Telephone Network
PtMP	Point to Multipoint
PtP	Point-to-point
QoS	Quality of Service
RA	Risk Analysis
RACF	Resource and Admission Control Functions
RACS	Resource and Admission Control Subsystem
RAN	Radio Access Network
RAT	Radio Access Technology
RB	Reserve Bandwidth
RF	Radio Frequency
RFID	Radio Frequency Identity
RLC	Radio Link Control
RoHS	Restriction of Hazardous Substances Directive
ROM	Read Only Memory
RRC	Radio Resource Control
RRM	Radio Resource Management

RSVP	Resource Reservation Protocol
RTCP	Real-Time Control Protocol
RTP	Real-Time Protocol
RTSP	Real-Time Streaming Protocol
SA	Social Accounting
SAE	System Architecture Evolution
SAN	Ship Area Network
SC	Score Card
SCF	Service Capability Function
SC-FDMA	Single Carrier-Frequency Division Multiple Access
SCS	Service Capability Server
SDR	Software Defined Radio
SDTV	Standard Definition Television
SEAAR	Social and Ethical Accounting, Auditing and Reporting
SGF	Signalling Gateway Function
SGSN	Serving GPRS Service Node
SID	Shared Information and Data model
SIM	Subscriber Identity Module
SIP	Session Initiation rotocol
SL	Service Level
SLA	Service Level Agreement
SLF	Subscription Locator Function
SM	Stakeholder Management
SMS	Short Message Service
SOA	Service-Oriented Architecture
SON	Self-Organizing Networks
SR	Social Responsibility
SW	Software
TA	Technology Assessment
TAN	Train Area Network
TBL	Triple Bottom Line
TCP	Transmission Control Protocol
TDM	Time Division Multiplexing
TETRA	Terrestrial Trunked Radio
TIA	Telecommunication Industry Association (NAFTA countries: USA, Canada, Mexico)
TISPAN-NGN	Telecommunication and Internet converged Services and Protocols for Advanced Networking-NGN
TMF	Telecom Management Forum
TNA	Technology Neutral Architecture
TOM	Telecommunication Operation Map
TOS	Type Of Service
TTA	Telecommunications Technology Association (Korea)
TTC	Telecommunication Technology Committee (Japan)
TTI	Transmission Time Interval
UDP	User Datagram Protocol
UE	User Equipment
UMB	Ultra Mobile Broadband

UML	Unified Modelling Language
UMTS	Universal Mobile Telecommunucation System
UMTS-S	UMTS Satellite component
UNI	User Network Interface
UPE	User Plane Entity
UPSF	User Profile Server Functions
URL	Uniform Resource Locator
UTRA	Universal Terrestrial Radio Access
UTRAN	Universal Terrestrial adio Access Network
UWB	Ultra Wide Band
VAN	Vehicle Area Network
VAS	Value-Added Service
VoIP	Voice over IP
WBCSD	World Business Council on Sustainable Development
WCDMA	Wideband CDMA
WiMAX	Worldwide interoperability for Microwave Access
WIN	Wireless Intelligent Network
WLAN	Wireless LAN
WWF	World Wildlife Fund
xDSL	x Digital Subscriber Line ($x =$ A, V, H, etc.)
xSIM	x Simple Input Method

Index